T0235909

Textbooks in Telecommunication Engineering

Series Editor
Tarek S. El-Bawab, PhD
Nile University
Giza, Egypt

Telecommunications have evolved to embrace almost all aspects of our everyday life, including education, research, health care, business, banking, entertainment, space, remote sensing, meteorology, defense, homeland security, and social media, among others. With such progress in Telecom, it became evident that specialized telecommunication engineering education programs are necessary to accelerate the pace of advancement in this field. These programs will focus on network science and engineering; have curricula, labs, and textbooks of their own; and should prepare future engineers and researchers for several emerging challenges. The IEEE Communications Society's Telecommunication Engineering Education (TEE) movement, led by Tarek S. El-Bawab, resulted in recognition of this field by the Accreditation Board for Engineering and Technology (ABET), November 1, 2014. The Springer's Series Textbooks in Telecommunication Engineering capitalizes on this milestone, and aims at designing, developing, and promoting high-quality textbooks to fulfill the teaching and research needs of this discipline, and those of related university curricula. The goal is to do so at both the undergraduate and graduate levels, and globally. The new series will supplement today's literature with modern and innovative telecommunication engineering textbooks and will make inroads in areas of network science and engineering where textbooks have been largely missing. The series aims at producing high-quality volumes featuring interactive content; innovative presentation media; classroom materials for students and professors; and dedicated websites. Book proposals are solicited in all topics of telecommunication engineering including, but not limited to: network architecture and protocols; traffic engineering; telecommunication signaling and control; network availability, reliability, protection, and restoration; network management; network security; network design, measurements, and modeling; broadband access; MSO/cable networks; VoIP and IPTV; transmission media and systems; switching and routing (from legacy to next-generation paradigms); telecommunication software; wireless communication systems; wireless, cellular and personal networks; satellite and space communications and networks; optical communications and networks; free-space optical communications; cognitive communications and networks; green communications and networks; heterogeneous networks; dynamic networks; storage networks; ad hoc and sensor networks; social networks; software defined networks; interactive and multimedia communications and networks; network applications and services; e-health; e-business; big data; Internet of things; telecom economics and business; telecom regulation and standardization; and telecommunication labs of all kinds. Proposals of interest should suggest textbooks that can be used to design university courses, either in full or in part. They should focus on recent advances in the field while capturing legacy principles that are necessary for students to understand the bases of the discipline and appreciate its evolution trends. Books in this series will provide high-quality illustrations, examples, problems and case studies. For further information, please contact: Dr. Tarek S. El-Bawab, Series Editor, Department of Electrical and Computer Engineering, Nile University, Egypt, telbawab@ieee.org; or Mary James, Senior Editor, Springer, mary.james@springer.com

More information about this series at http://www.springer.com/series/13835

Pramode Verma • Fan Zhang

The Economics of Telecommunication Services

An Engineering Perspective

 Springer

Pramode Verma
School of Electrical & Computer
Engineering
University of Oklahoma
Norman, OK, USA

Fan Zhang
Two-bit Beijing Technical
Beijing, China

Additional material to this book can be downloaded from (https://www.springer.com/us/book/9783030338640)

ISSN 2524-4345   ISSN 2524-4353   (electronic)
Textbooks in Telecommunication Engineering
ISBN 978-3-030-33867-1   ISBN 978-3-030-33865-7   (eBook)
https://doi.org/10.1007/978-3-030-33865-7

This Springer imprint is published by the registered company Springer Nature Switzerland AG.
The registered company address is: Gewerbestrasse 11, 6330 Cham, Switzerland

Preface

This book approaches the economics of telecommunication services from an engineering perspective. Telecommunication engineers are largely concerned with producing telecommunication centric goods and services that meet societal needs while providing an acceptable return to investors in the telecommunication industry. Networks are expected to serve traffic at a level of performance demanded by the user at an acceptable price point. The architecture of the technology deployed by the network are major determinants of these two user parameters. Both of these directly affect the economics of network services. Unquestionably, economics is an integral component in the business of offering network services.

Competition in the telecommunication history is relatively new. The monopoly environment of the past largely camouflaged the underlying interplay among technology, performance, and economics. Without the need for the resulting absence of a hard-core business based approach, pricing in the past was the outcome of such ambiguous and obscure parameters as "value of service" and "intended surplus of society." Although relatively recent, competition in the telecommunication industry is here to stay and exhibits itself both at the local and global levels. The primary intent of this book is to address the void that exists in understanding the economics of telecommunication networking from a technology perspective.

Telecommunication is a dynamic industry. This dynamic is driven by three major factors: technology, customer preferences, and the regulatory framework. Although the regulatory burden is largely in the "overhanging" mode, its impact will likely last for several decades even in the developed world. Each of these factors affects the economics of telecommunication services in different ways. This book explores the ways in which these factors will affect the evolution of the industry as a whole. Our approach is premised on the fact that, in order to be sustainable, pricing of telecommunication services will be driven by the underlying costs and we emphasize this approach throughout the book.

It is the belief of the authors that the economics of networks and network-based services warrants a fundamentally different approach to understanding its underlying cost and developing a cost-based pricing structure. There are three fundamental reasons for this. First, much of the classical economics applies to material goods rather than information based goods and services. Information manifests itself both as material goods and as a non-material entity defying the traditional approaches to characterizing its economic impact and potential. Second, information goods and services are driven, largely, by technology. Customer preferences do play a part of course, but it is at a much finer level, and more in form rather than in substance. In other words, the impact of technology influences the trajectory of the evolution of information goods and services to a far greater extent than customers' anticipation of and need for such evolution. The third reason is based on the fact that technology is now recognized as a factor of production, supplementing the traditional three factors of production labor, capital, and manpower. Unlike the other three, however, technology is known to have no limits. And within the realm of technology, information technology is, by far, the fastest growing component.

A notable barrier to motivation in understanding the economics of telecommunication-based goods and services in the past has been the fact that, until recently, it was treated as a regulated monopoly and owned, in most countries, by the state itself. Regulation of an industry, especially price regulation, distorts its pricing structure and any attempt to relate pricing to the underlying cost. Fortunately, a competitive landscape in the telecommunications space is emerging quickly. This will increasingly demand an understanding of the cost associated with telecommunication services.

The main emphasis in this book is on understanding the macroscopic parameters that affect the quality of service, capacity, and scalability along with the underlying cost of circuit and packet switched networks that offer a range of heterogeneous services with potentially varying performance parameters. Quality of service parameters that characterize circuit and packet switched networks from a macroscopic perspective are the grade of service and latency, respectively. Delivered traffic within the user's measure of acceptable quality of service is what the customer would pay for. The network service provider needs to ensure that the resources provisioned in the network are consistent with the capital and operational expenses it intends to incur in providing the service. Absent this consistency, the service provider will either cede its customers to competitors or go bankrupt. Indeed, this phenomenon was widespread during the telecommunications debacle some 15 years ago.

The pricing model proposed in this book is based on the cost of displaced opportunity as opposed to the cost of the elements of the network engaged in delivering a particular service. The displaced opportunity is characterized by the revenue associated with the service that the network could have alternatively delivered most efficiently using an identical level of its resources.

This book also introduces the use of game theory in pricing services in a competitive marketplace. Our belief is that as telecommunication transitions from the realm of regulated monopolies to competitive businesses, game theory is going to play an increasingly important role in the pricing of telecommunication services. This will be especially true in a multi-vendor environment where each vendor offers a number of services, each vendor's services supported by its common transport structure from which all its services are derived. The book has also used game theory to balance the interests of the customer, the service provider, and the regulator.

The book is suitable for use by the senior undergraduate, graduate students of telecommunications engineering, researchers, and practitioners in telecommunication engineering. However, it is primarily aimed at the practicing telecommunication engineer who is tasked with pricing telecommunication services in a competitive environment. It is the authors' hope that the book will bridge the gap between the science of economics as practiced by economists and the practice of pricing as carried out by network service providers.

Norman, OK, USA   Pramode Verma
Beijing, China   Fan Zhang
February 2020

Acknowledgements

This book is the outcome of a graduate course in telecommunications engineering that one of the authors (Pramode Verma) delivered over a decade at the University of Oklahoma-Tulsa. It has also incorporated results from research that his team conducted over the same period. Several students in the team received their master's and doctoral degrees based on their research. Part of the outcome of their research has been incorporated at several points in the book. More specifically, the authors would like to acknowledge the work of Drs. Mostafa Dahshan, Yingzhen Qu, Ziping Hu, and Ling Wang.

Pramode Verma would like to thank his wife Gita for her unwavering support over the past five decades. More specifically, Gita single-handedly assumed the burden of our physical relocation while the book was work in progress.

Fan Zhang would like to thank her husband Xuelin Li, for his help, long-time support, and encouragement throughout these years. Fan and Xuelin's daughter, Yifan, born during the preparation of this book, cooperated toward finishing it befitting her stature.

Contents

About the Authors

Pramode Verma is Professor Emeritus of Electrical and Computer Engineering at the Gallogly College of Engineering of the University of Oklahoma. Prior to that (1999–2016), he was Professor, Williams Chair in Telecommunications Networking, and Director of the Telecommunications Engineering Program. Before joining the University of Oklahoma, over a period of twenty-five years, he held a variety of professional, managerial, and leadership positions in the telecommunications industry, most notably at AT&T Bell Laboratories and Lucent Technologies. He has authored/co-authored several books and over 150 journal articles and conference papers and is the co-inventor of eleven patents. He has been a keynote speaker at several international conferences and conducted several workshops.

He holds a bachelor's degree in physics from Patna University, a bachelor's degree in engineering from the Indian Institute of Science, and a doctorate in engineering from Concordia University. He also holds an MBA from the Wharton School of the University of Pennsylvania.

Fan Zhang is Engineering Director at Two-bit Beijing Technical Company in China. Prior to joining Two-bit, she worked at Oracle, Amazon, and Cyngn (a start-up company) in Seattle, WA. Her research interests include pricing in networks and scalable fault-tolerant distributed systems. Her research has appeared in Netnomics, IET Communications, and presented at several international conferences. She obtained her doctorate in electrical and computer engineering from the University of Oklahoma, a bachelor's degree in management information system from Beijing University of Posts and Telecommunications (BUPT), and a master's degree in Telecommunications from BUPT.

Chapter 1
Characteristics and Characterization of Information Networks

1.1 What Is Information?

The contemporary society has been aptly named an Information Society. Our ability to transform, communicate, and act on information has made dramatic changes, especially over the last quarter century. These changes have affected the manner in which we conduct businesses, lead our personal lives, and run government operations, in fundamental ways. We anticipate even more changes in the future, possibly at an increasingly rapid rate.

But what is information? Information occurs when uncertainty is removed. Mathematically, we characterize information by the magnitude with which uncertainty has been removed. As a simple example, if the outcome—head or tail—resulting from the toss of a coin is known, we have secured one bit of information. Similarly, if one end of a transmission line transmits a 0 or a 1 and it is received at the other end correctly, the transmission line has conveyed one bit of information. The assumption, of course, is that both instances of the bit, namely, the 0 or the 1 are equally likely to occur. If this were not the case, the information received will be less than one bit. In the extreme case, if the receiving end had a priori knowledge of the bit (to be received), the information received would be zero bits.

We might ask ourselves if information is a tangible commodity or an abstract concept. Information behaves most of the time like a tangible commodity. It can be bartered or sold like a commodity. The information recorded in this book is priced by the publisher. The price might be different depending on whether the book is made available in an electronic format downloadable from the publisher's website or be in a paperback or hardcover edition.

Information can substitute for material goods in definitive ways and its value can be assessed as a material goods. For example, using better signaling techniques, railroads can carry more passenger and/or goods saving the cost of steel or labor in laying another track. Teleconferencing techniques, similarly, can displace the cost of travel and lodging while saving time for participants at the same time. The resulting

© Springer Nature Switzerland AG 2020
P. Verma, F. Zhang, *The Economics of Telecommunication Services*, Textbooks in Telecommunication Engineering, https://doi.org/10.1007/978-3-030-33865-7_1

savings are measurable, so the cost associated with the transport of information can be justified using economics metrics.

Information, however, is different from material goods in many ways. For example, it can be easily duplicated or stored at little cost. If information is stolen, there is no specific marker that would indicate that it is somehow diminished. This statement is generally, but not entirely, correct. For example, quantum cryptography is based on the fact that, for information coded in quantum bits, it is possible to detect if it has been observed or measured by a third party. This is quite unlike a situation where coins of gold removed from a safe could provide evidence of thievery. In a situation where a competitive bid from a vendor A is known to its competitor before the latter puts in its bid, vendor A may never be aware of the loss, yet its compromise could have a material impact. In addition to ease of duplication, even more compelling is the fact that the cost of distributing information has fallen down dramatically over the past few decades, and is likely to continue in the future.

Information has more value when it is made available at the right place at the right time. Telecommunication networks offer the vehicle over which information travels and reaches its intended destination. The management of telecommunication networks from an economic perspective is the primary intent of this book. The emphasis is on the economics of telecommunications networking.

1.2 Modalities of Information

Information may exist in more than one format. Speech, data, and moving or still images are examples of information in different formats. The characterization of information in each of these formats could be entirely different even though the underlying entity, namely, the bit, is the same. This might require some further elaboration.

Human speech, at its origin, exists in an analog format. It exhibits the singular characteristics of the specific speaker in a variety of ways. Retaining the tonal quality, the dynamic range of the speech, and the pauses or the gaps between words and syllables with fidelity across a network that serves as the medium between the two communicating parties is a large task. If these requirements are overlooked, human conversation will lose its subjective impact. Pauses and gaps between words or syllables, if not communicated with fidelity, might create a wrong impact on the listener. The resources needed in the network that would convey the "naturalness" associated with speech are very different from, say, conveying a transcript of the conversation written in words and transported as data between two human beings. A major objective of this book is to highlight the impact of resources needed to convey information with varying parameters of performance across a telecommunication network.

Speech is, most generally interactive and has, therefore, requirements related to the absolute delay it can tolerate without noticeable degradation. Additionally, stochastically varying delay or jitter is another requirement that affects the quality of conversation. Even though the frequency range of speech might be limited, the corresponding bandwidth requirement, when it is transported over a common user network, in particular, a packet switched network, could be very different depending on the degree to which the naturalness, as measured by jitter, is required to be communicated. The exploration of the amount of resources required in a network in communicating information to varying levels of quality is an important objective of this book. In other words, we will address the notion of quality from a subjective perspective while addressing its impact on the resources within the network in a quantifiable manner.

1.3 Analog and Digital Information

Information may be generated by analog or digital sources. Information originating from an analog source is represented as a continuous function of time. And its amplitude or intensity (depending on whether we are measuring the corresponding voltage or energy) itself varies continuously rather than in discrete steps. A digital source, on the other hand, is represented as a source that generates a sequence of symbols drawn from a finite set of alphabets. Most generally, the occurrence of the symbols takes place at discrete instants of time. This is not a hard and fast requirement, however, since it is possible that symbols originate with time as a continuum.

Despite the innate nature—analog or digital—of the information at its point of origin, it can be converted to the other format without losing any of its characteristics. Analog information can be converted into a digital format and vice versa. Independent of the modality of information at its point of origin or delivery, it is almost always transported in digital format. The digital format allows regeneration of symbols at appropriate points along the transport path thus ensuring that transmission errors do not accumulate during the passage of information in a network.

The contemporary network transports information in the digital format. The network is thus a medium for transporting binary digits between sources of information and intended destinations. This was not always the case, of course. In the history of telecommunication networking, the advent of digital networks is relatively recent. If we exclude telegraphy, basic building blocks of the first modern digital network appeared in the early 1960's as the T1 transmission system and was deployed by AT&T in its metropolitan areas providing trunks between telephone central offices [1]. When it comes to transporting information, a digitized network is preferable to its analog counterpart. Digital information can be regenerated rather than amplified thus preventing the accumulation of noise in the signal as it transitions a medium. A regenerator detects a symbol and then regenerates it for the next span of transmission.

From a graph theoretic standpoint, a network is simply a collection of nodes interconnected by transmission facilities. The function of the network is to facilitate communication between and among a pair (or more) of entities. From a general perspective, communication between two entities can be defined as a programmed response to a stimulus originating from one entity and conveyed to an appropriate destination. Communication thus includes processing and storage in addition to the commonly understood function of transmission.

As stated in Sect. 1.3, information in any mode can be converted into digital format. Data is defined as machine originated information in digital format. Speech transported over the network exists in digital format; however, it does not qualify as data. This distinction is important from the network's perspective because two ends transmitting and receiving voice signals may lose synchronization for, say, 40 ms without noticeably degraded performance from the users' perspective. Such a loss of communication will be unacceptable if data were transferred, however. In this case, either the network or the endpoints must be capable of recovering the lost data by other means, such as retransmission.

1.4 Networks: The Externality Factor

This section primarily addresses the impact of the size of the network on its economics. Stated differently, to what extent, if any, does the classical rule of economics—economy of scale—apply to communication networks. For the purposes of this section, the structure of the network, the performance parameters associated with transmission links, and other parameters related to network elements are entirely omitted. The analysis presented below is from a comparative standpoint only. It is not our intent here to compute the cost vs. size characterization of a network in absolute terms.

Network externality implies that the value of a network is higher if there are more users attached to it. A larger number of users increases the access space of each individual user. Metcalfe's Law [2], which is named after the inventor of Ethernet, states that the value of a network is quadratically related to the number of its users. This law is based on the observation that in a network with n users, each user can make $n-1$ other connections. This will result in a total of $n(n-1)$ possible associations. The total value of the network is thus proportional to $n(n-1)$, that is, approximately, n^2. This notion gives a large user base a competitive advantage for a network because each of the large network's users can communicate with a greater number of other users. The Metcalfe's Law obviously values each connection as being of equal value to the network as a whole. A newly added node (or customer), in order words, offers an equal benefit (of communication) to all the existing network nodes.

The Metcalfe's Law is an empirical observation made and heuristically validated during the frenzy associated with the explosion in the telecommunications industry fueled by deregulation in the mid-1990s. During this period, also characterized as the Internet bubble period, entrepreneurs, venture capitalists, and engineers believed in the steady and speedy commercial growth of the Internet. The Metcalfe's Law gave a quantitative explanation for the Internet boom with reasons like, "first move advantage," "Internet time," and "network effects." At that time, many companies invested heavily in new fiber infrastructures not only at a backbone level but also at the metropolitan level. Dense Wavelength Division Multiplexing made it possible to transport up to 160 light waves on a single strand of fiber with a combined bit rate into the range of terabits per second [4]. Ethernet technologies and the Internet Protocol made the connectivity services to be supported on the fiber infrastructure very inexpensive; a fact that likely is responsible for an overinvestment in the telecommunication infrastructure during the 1990s [5].

Metcalfe's law assigns equal value to all connections. This is a flaw because not all possible connections are equally valuable for each user. For example, in a large network such as the Internet, there are millions of potential connections between users. In general, connections are not all used with the same intensity, and most of them are not used at all. As a result, assigning equal value to all connection is not justified and a revision of Metcalfe's law is proposed in [6], based on the assumption that not all connections are equally valuable to the users. The assignment of value to each connection is based on the ZIPF's Law [3] which is an empirical rule. It says, for example, that in a long English language text, the most popular words, or the most used letters, are roughly related to frequencies of occurrence as: 1, 1/2, 1/3, ... This logic can be extended to a communication network with n users. For each user, the value of connections to other users will be proportional to $1 + 1/2 + 1/3 + \cdots + 1/n$, which approaches roughly $\ln n$. There are other $n - 1$ users who get similar value from the network and the value of the network is thus proportional to $n \ln n$ in the revised Metcalfe's Law.

The revised Metcalfe's law shows that the value of the network grows faster than its size in linear terms and has a form of $n \ln n$. The $n \ln n$ valuation describes a slower growth (than what the classical Metcalfe's Law would predict) in the value of dot-com companies, and explains the Internet bubble from another angle.

In this section, we have reviewed the network externalities: The value of the network increases as $n \ln(n)$, where n is its user base. This is faster than the linear growth while the cost of the network is, at most, linear to its user base. A network provider has, thus, a great incentive to price its services attractively to increase the user base.

1.4.1 Network Mergers

Consider two networks: network N_1 and Network N_2 with user bases of n_1 and n_2. Using the (classical) Metcalfe's Law, let the parameters $V N_1$ and $V N_2$ capture their respective values. We can write, $V N_1 = k n_1^2$, and $V N_2 = k n_2^2$, where k is a constant.

The combined value of the networks as separate entities can be given as:

$$V(N_1 + N_2) = k(n_1^2 + n_2^2) \tag{1.1}$$

The combined value of the merged network N will be,

$$VN = k(n_1 + n_2)^2 \qquad (1.2)$$

From Eqs. (1.1) and (1.2), it can be easily seen that, irrespective of the values of n_1 and n_2, the value of the combined network is always higher. This should offer motivation to networks of any (relative) sizes to merge. However, in practice, networks with a large user base have not been observed to court their much smaller competitors with the intent of a possible merger. In the following sub-section, we examine if the relative values of n_1 and n_2 affect the motivation for merger, assuming that $n_1 + n_2$ is a constant.

1.4.2 Motivation for Merging Networks

As in the preceding section, let there be two networks, N_1 and N_2, with user bases of n_1 and n_2, respectively. A combined network N will have a user base of $n = n_1 + n_2$. We wish to answer the following question: If n were to remain constant, what is the relationship between n_1 and n_2 that will maximize the value of N? In the following analyses, we address this question for both the classical and the modified Metcalfe's Law.

First, using the classical Metcalfe's Law, we note that the value V_1 of the network N_1 is given as, $VN_1 = kn_1^2$, where k is a constant. Similarly, $VN_2 = kn_2^2$.

The value of the combined network N can be given as, $VN = k(n_1 + n_2)^2$. The additional value created due to merger is,

$$v = 2n_1 n_2 = 2n_1(N - n_1) \qquad (1.3)$$

A point of inflection in Eq. (1.3) would occur when, $\frac{\partial v}{\partial n_1} = 0$, i.e., when, $n_1 = N/2$.

It can be easily seen that $\frac{\partial^2 v}{\partial n_1^2}$ is negative proving that the point of inflection corresponds to a maxima. We note that the networks we considered have placed identical value on each customer, both in terms of cost of provisioning and potential for revenue. This is not always the case, of course.

We have shown above that combination of networks always yields an advantage for the merged entity. This advantage can be quantified for each of the merged entities as follows. The value of the merged entity increases by $2n_1 n_2$ over and above the sum of the values of the individual networks. It is reasonable to assume that the added value, namely, $2kn_1 n_2$ is attributed equally to the two merged networks. With this assumption, the value of the network N_1 increases by $kn_1 n_2$ and becomes $kn_1(n_1 + n_2)$. Similarly, the value of the network N_2 increases to $kn_2(n_1 + n_2)$. Obviously, the relative increase in the value of the smaller network will be much higher than that of the larger network. This would explain the motivation of smaller networks to merge with larger entities.

The analysis can be extented to include the modified Metcalfe's Law as follows.

Following the same notations as before, the sum of the values of the two networks prior to merging is: $k(n_1 \ln n_1 + n_2 \ln n_2)$. Similarly the value of the combined network is: $kn \ln n$

It can be easily seen that,

$$n \ln n > n_1 \ln n_1 + n_2 \ln n_2 \tag{1.4}$$

since

$$n \ln n = n_1 \ln n + n_2 \ln n \tag{1.5}$$

Hence the merger benefits combining two networks accrues under the modified Metcalfe's Law as well.

In order to show that optimal value of merging is realized when the merging entities have equal value, we proceed as follows.

Following the same approach as before, an optimum will occur when,

$$\frac{\partial[n_1 \ln n_1 + n_2 \ln n_2]}{\partial n_1} = 0 \tag{1.6}$$

or

$$\frac{\partial[n_1 \ln n_1 + (n - n_1) \ln(n - n_1)]}{\partial n_1} = 0 \tag{1.7}$$

Evaluating each of the terms separately,

$$\frac{\partial[n_1 \ln n_1]}{\partial n_1} = \ln n_1 + 1 \tag{1.8}$$

since (See Appendix A),

$$\frac{\partial[\ln n]}{\partial n} = \frac{1}{n} \tag{1.9}$$

and,

$$\frac{\partial[(n - n_1) \ln(n - n_1)]}{\partial n_1} = -[1 + \ln(n - n_1)] \tag{1.10}$$

From (1.8) (1.10), and (1.7) can be rewritten as,

$$\ln \frac{n_1}{n - n_1} = 0 \tag{1.11}$$

or $n_1 = n/2$ which proves that the maximum value of merger is realized when the merging entities have equal value.

1.5   Merger of Heterogeneous Networks

Underlying the motivation for merger of two separate networks, discussed in the last section, was the fact that each user presented a uniform demand on the network and contributed an identical value to the network. Merger of heterogenous networks is a more complex issue from an analytical standpoint. We address this question in Appendix B.

1.6   Illustrative Example

Problem Two networks with numbers of subscribers equal to 2000 and 3000, respectively, merge. Find the relative increase in their respective values using (a) the original Metcalfe's law, and (b) its enhanced version.

Solution (a)
Prior to Merger

- Value of network1 $= k2000^2$
- Value of network2 $= k4000^2$

After Merger

- Value of the combined network $= k6000^2$
- Total *relative* value increase $= \frac{6000^2}{2000^2 + 4000^2} = 180\%$
- Total increase in value $= k(6000^2 - 2000^2 - 4000^2) = 1.6k * 10^7$

We can allocate the value increase equally between the two networks, or distribute it between the two networks in proportion to their respective number of subscribers prior to merger. Assuming we follow the value distribution proportionality, since network1 had an initial value equal to a quarter of the value of network2, value increase of network1 $= 1/5$ of the total increase in value while network2 gains a value increase of 4/5 of the total increase in value.

If the increased value were to be awarded equally between the two networks, each network would gain a value of $8k * 10^6$.

Solution (b)
Prior to Merger

- Value of network1 $= k2000 \ln 2000 = k15201.8$
- Value of network2 $= k4000 \ln 4000 = k33176.2$

After Merger

- Value of the combined network $= k6000\ln 6000 = k52197.1$
- Total *relative* value increase $= \frac{52197.1}{15201.8+33176.2} = 108\%$
- Total increase in value: $k(52197.1 - 15201.8 - 33176.2) = k3819.1$

Distributing the value increase in proportion to their original values, value increase of network1 $= k\frac{15201.8*3819.1}{15201.8+33176.2}$ or $k1200.1$. Network2 gains a value of $k2619$.

If the increase in value were to be awarded equally, each network would gain a value of $k1909.55$.

Problems

1.1 State Metcalfe's Law. How does it relate the value of a network to the number of subscribers it serves? Give one reason that would lead you to suspect that it might be too aggressive.

1.2 Two networks with numbers of subscribers equal to 2000 and 3000, respectively, merge. Find the relative increase in their respective values using the original Metcalfe's Law and its revised version.

1.3 The naturalness of voice in telephony is governed by two parameters—fidelity and dynamic range. Which elements of the PCM system govern these two parameters? How do we arrive at 1.544 Mbps as the line rate of the T1 system? Explain the use of the framing bit in a T1 frame. *(This question is not reviewed in the book. However, it will be a good exercise to address it in the class room or have students look up its answer in any book on telecommunication.)*

1.4 Cable television channels today are priced in bundles. At the same time, it is well known that most subscribers watch just a few channels in the bundle, typically just about 10–20 channels in a 150-channel package. In a recent survey, most consumers (>90%) expressed their desire to pay for only those channels they regularly watch. It is well known that the popularity of the different channels is vastly different, and these channels receive different amounts from the cable service provider depending on the eye balls they attract. The Internet-based service pricing for a single channel (e.g., Netflix) is about \$10/month. State the likely consequences of a la carte pricing of cable channels on the consumers as well as the service providers. What would happen to the channels that are generally unpopular with the masses? *(This question is not reviewed in the book. However, it will be good exercise to address it in the class room.)*

References

1. M.R. Aaron, PCM Transmission in the exchange plant. Bell Syst. Tech. J. **41**(1), 90–141 (1962)
2. C. Shapiro, H.R. Varian, *Information Rules* (Harvard Business Press, Cambridge, 1999), p. 184
3. C.D. Manning, H. Schütze, *Foundations of Statistical Natural Language Processing* (MIT Press, New York, 1999), pp. 24
4. NTT(2010-03-25), *World Record 69-Terabit Capacity for Optical Transmission over a Single Optical Fiber, Press release* (2010)
5. C. Courcoubetis, R. Weber, *Pricing Communication Networks: Economics, Technology, and Modelling, West Sussex, England* (Wiley, Hoboken, 2003)
6. B. Briscoe, A. Odlyzko, B. Tilly, Metcalfe's law is wrong—communications networks increase in value as they add members-but by how much? IEEE Spectr. **43**(7), 34–39 (2006)

Chapter 2
Drivers of the Telecommunication Industry

2.1 Bandwidth

Bandwidth or transmission capacity is one of the two fundamental resources for information transfer from source to destination. The source and the destination are connected through a medium or a combination of media such as a copper or a fiber-optic cable, or free space. The other fundamental resource, discussed in Sect. 2.2, is processing power. Information can originate in analog or digital format and it can be transported over the medium or media in either of these formats.

Analog information sources generate information that is a continuous function of time and can assume a continuous range of values. Digital information, on the other hand, is represented by discrete symbols or alphabets from a finite set. Digital information, generally speaking, originates and is transmitted at discrete instants of time. However, this is not absolutely essential. For example, a black and white image scan can assume only two discrete values which can occur as a continuous function of time (or space).

Depending upon the distance between the source and the destination, the intervening medium can include a repeater which will compensate for the loss of signal as it deteriorates with distance. We emphasize the fact that the transmission medium is not inherently analog or digital. If the repeater is an amplifier which compensates for the loss, the transmission system is analog. On the other hand, if the repeater regenerates the signal, the transmission system is digital. If the transmission system is digital, an analog signal can be transmitted over it by first converting it into a digital format. Information in analog format can be converted into digital format without any loss of information and vice versa. Because digital information can be regenerated at any point and, in the process, remove any signal deterioration, it is the preferred mode of transmission, especially over long distances.

Analog bandwidth is measured in Hertz, abbreviated as Hz, and digital bandwidth in terms of bits per second or b/s. Digital signal processing technology has considerably improved the spectral efficiency of transmission. Spectral efficiency is

© Springer Nature Switzerland AG 2020

P. Verma, F. Zhang, *The Economics of Telecommunication Services*, Textbooks in Telecommunication Engineering, https://doi.org/10.1007/978-3-030-33865-7_2

simply the digital bandwidth in terms of bits per second that can be squeezed into one Hertz of analog bandwidth. Spectral efficiency is most important in wireless communication where the available analog bandwidth in a given space is limited and cannot be duplicated as is the case of communication in a guided medium, such as a fiber-optic cable. In the latter, multiple cables can be laid side by side increasing the bandwidth proportionately.

The analog bandwidth and the digital carrying capacity of a transmission channel are related by the classical equation developed by Claude Shannon in 1948 [1] and is expressed as follows:

$$C = W \log_2 \left(1 + \frac{S}{N} \right) \qquad (2.1)$$

where C is the channel capacity in bits per second, W is the bandwidth in Hertz, and S and N are the power levels associated with the signal and the noise, respectively.

Equation (2.1) represents an upper bound of channel capacity under ideal conditions. If we consider an analog telephone channel with the following bandwidth and signal-to-noise characteristics: $W = 3000$ Hz, $S/N = 30$ dB or 10^3 then, $C = 3000 \log_2(1 + 1000) = 29,880$ b/s.

Capacity of optical fibers has dramatically improved over time. Current technology for commercial applications allows each fiber to have several independent transmission channels with each channel having a transmission capacity of up to 40 Mb/s. Dense wavelength division multiplexing techniques can lead to each fiber having several tens of independent channels allowing a single fiber to carry up to a few Tb/s of information. The spectral efficiency has correspondingly grown from about 1 bit/s/Hz to 10 bits/s/Hz over the past 15 years.

Needless to say, the dramatic improvement in the carrying capacity of fiber-optic technology has brought down the transmission cost by orders of magnitude over the corresponding period. Without such a reduction, the fast evolving cloud technology allowing information to be stored and processed remotely and made available on demand anywhere would not be economical.

2.2 Processing Power

Much of the evolution of telecommunications networking can be directly attributed to digital signal processing techniques. The processing technology based on semiconductors has gone through revolutionary changes over the last 40 years. This change is often captured as the Moore's Law, which is an empirical observation made by Gordon Moore in 1965 [2]. The economic impact of Moore's Law is a fundamental driver for the reduction in costs of telecommunication technology. From an economic perspective, the law states that the cost of processing power would continue to halve every 12 months in the foreseeable future. This observation has held since its inception and has driven the demand for telecommunication services correspondingly. The positive outlook for the chip industry continues unabated till now. Sometime in 2017, semiconductor manufacturers shipped processors using the 10-nm chip-manufacturing technology [3].

We note in passing that the computing industry as a whole appears to be driven by the Moore's Law. As a result, anticipating the increase in processing power, the application writers continue to create applications that are ever more functional and that would depend on the anticipated increase in processing power. The technology layers in between the raw processing power and the application, the operating system, software languages, firmware, and middleware, continue to be driven accordingly. This virtual lockstep among the different facets of the computing industry has advanced the information industry as a whole to new heights where the end user is looking for new applications and devices at regular intervals. This is a unique phenomenon in the computing and telecommunication world, and is truly global in scope.

Processing power along with the transmission capacity constitute the raw material for telecommunication networking. A continuing reduction in their costs would imply that telecommunication networking would displace other components of business or personal endeavors to the extent that a cross elasticity exists between these components and telecommunication-based applications. Indeed, this has been the case over the past several decades. While forecasting of the evolution of technology is fraught with risks, we do posit that the combination of reducing costs of bandwidth and processing technologies, when combined with human ingenuity in conceiving novel applications, will continue to enhance the relevance of telecommunication networking in our personal and business world. Businesses would do well by recognizing this trend and plan for correspondingly enhanced investment in their telecommunication networking needs.

2.3 Composition of the Telecommunication Industry

Like any other industry, the telecommunication industry has two major segments: the customer segment and the supplier segment. Customers are further segmented into business customers and consumers. Each of these segments buys products and services. The supplier segment is further segmented into service providers and equipment providers. Each of these sub-segments offers services or equipment to both business customers and consumers. The segments are illustrated in Fig. 2.1. As shown, the fact that the same service provider provides telecommunication service to both the retail consumer and the business enterprise implies that there is an inherent economy of scale associated with telecommunication services. Likewise for equipment providers.

Innate characteristics of telecommunication products and services suppliers are that they (suppliers) are global in scope as far as the footprints of their offerings are concerned. The growing mobility of people in both the personal space and

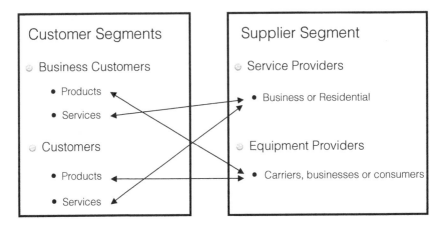

Fig. 2.1 Composition of the telecommunication industry

the business space would imply that planning for launching telecommunication products and services should consider the global scope of the business.

As shown in Fig. 2.1, service providers offer services to both the business and retail customers. Similarly, equipment providers supply equipment to both the customer segments. Suppliers in the telecommunication business, in other words, do not differentiate themselves by specializing in a specific market segment characterized as the consumer or the enterprise market.

2.4   Telecommunication Services Utility and Demand

Sections 2.1 and 2.2 have captured the fundamental drivers of telecommunication services. These factors are based on the increasing capability of the underlying technology while at the same time reducing its costs. Another important factor driving the demand for telecommunication services is the fast pace at which user friendly applications are emerging. The combination of these two factors is responsible, in large part, for expansion of the telecommunication services market.

This section captures the relationship between the value of telecommunication services as perceived by the customer and the price paid for it. Economists present the rationale for customer i buying a certain amount of a specific service by its utility function, u_i. If x is the amount of a service and the perceived value of the service by the customer i is v_i, then the utility of the service purchased is represented by,

$$u_i(x) = v_i(x) - px \tag{2.2}$$

Equation (2.2) assumes a linear relationship between the unit price paid by the customer and x which represents the amount of service purchased by the customer. The price associated with one unit of service is represented by p.

In telecommunication parlance, the amount of service may, for example, be represented by the maximum speed of download that a customer could get from the Internet using the access mechanism he or she pays for. For a wireless customer, x might represent the total number of bytes the customer transfers to or receives from the Internet over a period of time, say, a month. We should caution the reader that the total price paid is seldom a linear function of the amount of service purchased. Nevertheless, Eq. (2.2) offers an important insight into the dynamics between the amount of service a customer x would buy at a given unit price p associated with the service.

A general assumption economists make is that the amount of service x purchased by the customer i is consistent with maximization of its utility $u_i(x)$. For a given market of n customers buying k services, we can generalize Eq. (2.2) as:

$$u_i(x) = maximize(v_i(x) - p^T x) \tag{2.3}$$

where x and p are two vectors as $x = (x_1, x_2, \ldots, x_k)$; $p = (p_1, p_2, \ldots, p_k)$. $v_i(x)$ is the value to customer i of having a vector quantity of service x.

Let us consider the value associated with telecommunication services as a function of the amount of service purchased by a customer. In general, for a commodity, the value increases as a function of the amount of commodity purchased. This is true for telecommunication services as well. The function is seldom linear, however, except possibly in the case of specific commodities that are consumables, such as razor blades. The linear relationship would have other caveats as well. For a linear increase in the value of a commodity, for example, storage should have negligible cost, and the durability of the commodity purchased should be high. Figure 2.2 presents a hypothetical telecommunication service and plots the value of the service against the amount of service purchased. The shape of the curve is almost always concave signifying that the value increases as the amount of service purchased increases, but the corresponding increment in value reduces at the same time. For

Fig. 2.2 A concave valuation function

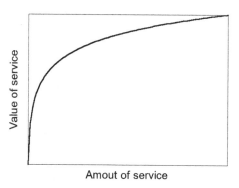

the purposes of this chapter, we will assume this to be always the case, i.e., v_i is strictly increasing, and strictly concave.

2.5 Price Elasticity of Demand

Economists use price elasticity to define the relationship between price and demand. In general, as the price of a commodity or service increases, its demand reduces, often by a large factor. Correspondingly, as the price reduces, the demand increases. Mathematically, the demand x for a service j denoted as x_j is related to its price p_j as:

$$\epsilon_j = (\nabla x_j / x_j)/(\nabla p_j / p_j) \tag{2.4}$$

Here ε_j is the price elasticity of service j: the percentage change in the demand per percentage change in price.

The price elasticity of demand for telecommunication services is substantively different from the price elasticity of common material goods and services. Let us consider the elasticity associated with the speed of accessing Internet services and its price. Up until about 1999, access to the Internet was through the PSTN limiting the speed of access to 9.6 kb/s. With the advent of ADSL (Asynchronous Digital Subscriber Line), this speed increased substantively, and continues to increase as more advanced access technologies emerge. Users continue to demand and pay for higher access speeds as innovative visual applications emerge. Any attempt to place a limit on the maximum speed that will suffice for a consumer is not yet in sight. In other words, the demand for higher access speeds shows no sign of abatement, even as the price for leasing a higher speed access mechanism continues to increase.

As an alternative to price elasticity of demand for telecommunication services, let us consider the price elasticity of demand for a transportation vehicle, such as a bicycle or a car. First, a user can use a transportation vehicle only one at a time. A number of transportation vehicles cannot be aggregated to provide a higher level of service to the user. This is very unlike a higher speed access mechanism which will offer increasing level of value to the customer, say, by enhancing the quality of a visual application or a reduction in the latency of an interactive service. As another example, to a speed-trader on the stock market, a lower latency compared to his or her competitors can offer an unlimited profit potential. There are no costs, other than the price of a low response time, incurred by the trader. On the other hand, in the case of transportation vehicles, there are other costs such as the cost of garaging a motor vehicle and costs associated with maintaining a license and insurance to keep the motor vehicle road worthy.

One might ask the question: Is there a maximum access speed above which a consumer would find no additional value? We'd hazard to guess there is not any. First, there is no known limit to the speed at which human minds can process information. And even if there were any, with the use of appropriate inanimate tools, this limit would be infinitely expandable. Suffice it to say that this limit would not be lower than the maximum speed with which information can be processed or transported.

There are many situations where goods or services of one kind have the ability to substitute for or, alternatively, increase the demand for another item. The substitutability or complementarity among goods or services is known as cross elasticity of demand. The substitutability of x by y will reduce the demand for x while increasing the demand for y, assuming the price associated with y to offset an equivalent amount of x is lower. Complementarity between x and y will increase the demand for both. For example, an increase in the number of cars sold will correspondingly increase the demand tires increasing the market for tire manufacturers. We discuss cross elasticity of demand between goods and services in Sect. 2.6.

2.6 Cross Elasticity of Demand

As mentioned in Sect. 2.5, it is well known that material goods and/or services of one kind have the potential to spur or suppress the demand for another. The cross elasticity ε_{jk} between the demand for two services j and k represented by x_j and x_k is defined as:

$$\epsilon_{jk} = (\nabla x_j / x_j)/(\nabla p_k / p_k) \tag{2.5}$$

From Eq. (2.5), if $\varepsilon_{jk} > 0$, it implies that service j and service k are substitutes; if $\varepsilon_{jk} = 0$; it implies that service j and service k are independent; if $\varepsilon_{jk} < 0$; it implies that service j and service k complement each other.

How does the pricing of telecommunication services depend on its price elasticity and what are the goods and services with which it has cross elasticity? We discuss it in the following section.

2.7 Price and Cross Elasticity of Demand for Telecommunication Services

In order to project the demand for telecommunication services, it is important to understand its price elasticity in isolation as well as in conjunction with other goods and services with which it has cross elasticity. On its own, as one example, the user would ask the following question: Is the increased speed of Internet download consistent with increased productivity that can be translated into a tangible benefit?

Similarly, in the case of cross elasticity, the question would be: Are telecommunication services elastic with other goods and services, seemingly unrelated? The answer, in general, is in the affirmative. As one example, consider the case of a plumber who, in the absence of cellular telephone services, was always obliged to drive to the office to pick up the work order that would be his next job. Before investing in cellular telephone service he would investigate if the increased cost of cellular service, and the necessary instrument, is lower than the additional revenue the plumber can realize by his increased ability to service more customers.

2.8   Summary

This chapter has identified the rationale for continuing growth of telecommunication services in the foreseeable future. The rationale is driven by two fundamental factors: reducing costs of the raw material necessary to implement telecommunication services, and the demand and cross elasticity of demand associated with telecommunication services. We have noted that, in the foreseeable future, the cost of bandwidth—both in guided media as well as in free space—will likely continue to decrease at the same rate as in the past decades. Similarly, the cost of processing power would continue to decrease at a rate projected by the Moore's Law which results in dramatic reduction in the cost of processing power over time.

   In terms of the ability of telecommunication services to offer increased value to the customer, the picture is equally bright. There appears to be an insatiable demand for bandwidth driven in large part due to enormous growth in visual services and ability to make these services instantly available at any place, any time. Similarly, telecommunication services continue to offer cross elasticity with both material goods and services. The combination of these would continue to propel the growth of telecommunication services in the foreseeable future. Furthermore, telecommunication services have the ability to displace other goods and services at a lower cost while offering a higher value.

Problems

2.1 An analog signal with 4 kHz bandwidth is sampled at 10000 times per second. Each sample is quantized into one of 256 equally likely events. Assume successive samples are independent. What is the information rate of the source?

2.2 The signal-to-noise ratio of a telecommunications channel is 30 dB. How would the transmission capacity of the channel in bits per second increase if the signal power increased three times? Assume the bandwidth of the channel in Hz and the noise power remain the same.

2.3 The signal-to-noise ratio of a telecommunication channel is varied from 30 to 40 dB. In what proportion does the carrying capacity of the link in bits per second change? Assume the bandwidth of the channel in Hz remains constant.

2.4 How is the telecommunication market segmented? What specific characteristics do you see in this market?

2.5 As shown in Eq. (2.2), an individual investor tries to maximize her or his utility function. On the other hand, the regulator tries to maximize the social welfares defined below for a specific case.

Suppose there is one producer, and a set of consumers, $N = 1, \ldots, n$. Let x_i denote the vector of quantities of k services consumed by customer i. The consumer surplus, $s = \sum_{i \in N} u_i(x_i) - c(x)$

For the regulator, maximization of s is the goal, subject to the total demand $x = x_1 + x_2 + \ldots + x_n$. In order to solve this problem as an illustration, we make use of the Lagrange Multiplier as follows:

$$L = \sum_{i \in N} u_i(x_i) - c(x) + p(x - x_1 - \ldots - x_n)$$

We can rewrite the above problem as the following two maximization problems:

- producer: $max_x(px - c(x))$
- consumer i: $max_{x_i}(u_i(x_i) - px_i)$

The following is a simple example illustrating the social welfare maximization problem. Suppose there is one producer and two consumers. The producer has two different services. We assume, for example, that the concave user utility function is $wlog(x)$ and the convex cost function is a fixed value c. We have

$$s = w_{11} \log x_{11} + w_{12} \log x_{12} + w_{21} \log x_{21} + w_{22} \log x_{22} - c$$

Assume that, $w_{11} = 10$, $w_{12} = 15$, $w_{21} = 10$, $w_{22} = 30$, $c = 15$, the total supply $x = [10, 10]$. Find the solution for this maximization problem.

References

1. C.E. Shannon, *The Mathematical Theory of Communication* (University of Illinois Press, Urbana, 1998) [1949]
2. G.E. Moore, *Cramming More Components Onto Integrated Circuits.* Electronics Magazine (1965-04-19). Retrieved 2018-07-01
3. IEEE Spectrum, pp.52–53 (2017)

Chapter 3
Graph Theoretic Characterization of Communication Networks

3.1 Networks and Graph Theory

The science of graph theory is deemed to have originated from the now legendary Königsberg bridge problem [1]. Königsberg, a city in eastern Prussia had seven bridges with four land masses as shown in Fig. 3.1. The city wanted a solution to the following problem: Does a path exist for someone to cross all the seven bridges without having to cross the same bridge twice? The famous mathematician Leonhard Euler in 1736 offered a solution to the problem by stating, and proving mathematically, that such a path does not exist. In the process, Euler also founded a new branch of mathematics called graph theory. Euler converted the topology of the Königsberg bridge to a graph with four nodes, one each for the four land masses, and seven links for each of the bridges. He also observed that a continuous path that goes through all the bridges can have only one starting and one exit point. Furthermore, nodes with an odd number of links must be either the starting or the end point of the travel path. The Königsberg bridge graph (also shown in Fig. 3.1) has four nodes with an odd number of links, making such a path impossible. A new bridge between nodes 2 and 4 would solve the problem, leaving only nodes 1 and 3 with an odd number of links. A path that will traverse all the bridges without crossing any bridge twice can now be easily found.

Figure 3.1 also shows a tabular representation of the graph known as the Adjacency Matrix. The adjacency matrix has four rows and four columns. Each row and column represents whether or not there is a link between a particular node to another. For example, node 1 is directly connected to nodes 2, 3, and 4 making the three entries in the first row, against columns 2, 3, and 4, a "1." The first entry in row 1 is zero. Similarly, the third row specifies that node 3 is directly connected to nodes 1, 2, and 4 but not to itself.

Graph theory has addressed and solved many issues that had puzzled mathematicians and scientists since Euler solved the Königsberg bridge problem in 1736.

© Springer Nature Switzerland AG 2020
P. Verma, F. Zhang, *The Economics of Telecommunication Services*, Textbooks in Telecommunication Engineering, https://doi.org/10.1007/978-3-030-33865-7_3

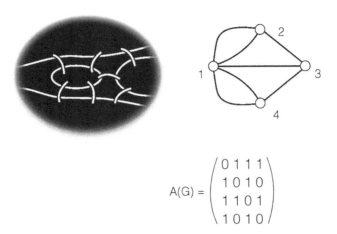

$$A(G) = \begin{pmatrix} 0 & 1 & 1 & 1 \\ 1 & 0 & 1 & 0 \\ 1 & 1 & 0 & 1 \\ 1 & 0 & 1 & 0 \end{pmatrix}$$

Fig. 3.1 The Königsberg bridge problem

We have defined communication networks as a set of nodes interconnected by transmission links. A graph of communication networks can thus be easily formulated. The graph of a network cannot capture all its properties. Even so, it offers such an important abstraction of the network that several characteristics of the network can be easily deciphered from the graph using well established results from graph theory.

3.2   Graphs of Some Well-Known Networks

Figure 3.2 represents graphs of some well-known network topologies. The network labeled Barbells simply connects two nodes by a transmission facility. The Line Network connects a number of points in tandem as shown. The topology of the Star Network has a "hub and spoke" architecture as shown. The Star Network is also a centralized network. The Ring Network is similar to a line network with the two ends connected by a transmission line. A random network has transmission links assigned between pairs of nodes chosen randomly.

Even a cursory look at the five topologies shown in Fig. 3.2 can result in the following observations. Loss of even a single transmission link in the Line Network will disconnect the network into two disjoint pieces. In the Star Network, failure of a single link will isolate only one node; the rest of the nodes will be able to communicate between and among themselves. However, if the central node is compromised, the entire network will become dysfunctional. For the Ring Network, assuming each link is bi-directional, failure of one of the links in any of the two one-way rings will still allow communication among all the nodes.

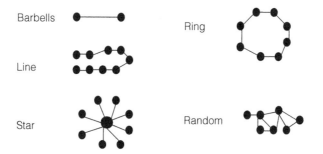

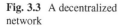

Fig. 3.2   Some well-known network topologies

Fig. 3.3   A decentralized
network

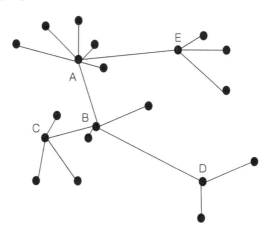

The public switched telecommunication network (PSTN) is the prime example of a telecommunication network. It is global in scope with the ability to connect any telephone to any other on demand without human intervention. The Internet, which is rapidly growing worldwide, is another prime example of a telecommunication network. The Internet has the ability to include a wide variety of end devices to communicate among themselves. Originally conceived as a functionally separate (from the PSTN) network, the geographic spread of the Internet with the ability to accommodate almost any device as an end point, combined with unlimited range of functionalities, its scope as a common user communication vehicle is truly unlimited.

There is a whole range of network topologies that can be analyzed using tools developed by graph theorists over more than two centuries. The networks include transportation networks, power grids, pipelines for distributing natural gas, the network of rivers, and social networks. Graph theory offers the tools that will analyze the topological impact of these networks on the functionalities they are intended to offer to the user.

Often, a real network is a decentralized network, shown in Fig. 3.3, which consists of a small number of star networks whose hubs are interconnected. Nodes

A, B, C, D, and E are hubs of the network. The rest of the nodes are connected to a hub with one link. If the hubs are fully interconnected (not shown in Fig. 3.3), communication from any node to another will involve no more than three links. This is because, in the fully interconnected case, a hub-to-hub communication will require just one link and each end node will have just one link to the respective hub.

The number of links involved in effecting end-to-end communication has a profound impact on the resources the network consumes and on the end-to-end performance as experienced by the user. The intent of a well-designed network is to have as few links as possible between the source and the destination nodes. Mathematical analysis of the impact of the number of links between the source and destination on performance experienced by the user is addressed in Chaps. 6 and 7 and the following chapters.

3.3 Networks of Information

Many of the examples of networks presented in Sect. 3.2 carry commodities such as natural gas or convey electrons such as in the power grid network. Networks of information, on the other hand, convey information or link documents for retrieval and use by the end user. One such information network is the www network which links information repertoires such as documents or information in different formats to each other or from a source to a sink using hyperlinks. The Internet accommodates all such linkages.

Other examples of information networks are citation networks, biological networks or neural networks. The key commodity that is transferred by these networks is information. Since information can be measured, the amount of information conveyed by a network vs. the amount of resources consumed by the network is an important measure of the efficacy of the network.

The transport of information is not the same as the transport of a commodity in the conventional sense and is qualified by "quality of service." The quality of service is a critical factor in the choice of a network that will carry the information. Information networks thus can be compared on the basis of the cost incurred by the network to convey information while meeting the quality of service needs of the application.

In the following section, the legacy PSTN (Public Switched Telecommunication Network) is addressed and its graphical representation shown.

3.4 The Legacy PSTN: A Graphical Representation

The legacy PSTN is a decentralized network where a path length between any pair of end points could have up to 9 links. This network is hierarchically organized with an incoming call going up the hierarchy from the Class 5 central office (which

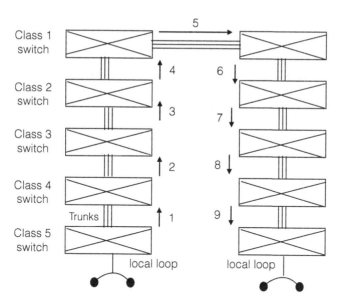

Fig. 3.4 Legacy PSTN hierarchy

terminates the calling or the called telephones) to a Class 4 central office (which acts as a tandem switch), and so on up to the Class 1 switch, and then descending toward the Class 5 switch on which the called party is terminated. Class 1 switches represent the highest level of hierarchy and are fully interconnected. A hypothetical hierarchy of a call that involves all possible nine links is shown in Fig. 3.4.

As stated in the last paragraph, the legacy network can include up to nine links between the calling and the called end points, excluding the local loops. This might sound like a lot of links. However, even the current Internet architecture allows several links on an end-to-end basis. The impact of the number of links on the performance perceived by the user is addressed in several succeeding chapters of this book. In brief, a large number of links degrades the performance drastically as the traffic volume increases.

We do note, however, that the legacy network reflected, somewhat optimally, the then engineering tradeoff between the cost of transmission and the cost of providing switching nodes. The cost of transmission has come down sharply relative to the cost of switching or routing over the last several decades. This has resulted in reducing the number of links between the two communicating entities and has favorably impacted the overall cost of transporting information while improving performance.

3.5   Graphical Representation of Networks

Section 3.1 has already introduced the graphical representation of a network. From a generalized perspective, a network is treated as a graph G, where

$$G = f(n, m, F) \tag{3.1}$$

n is the set of nodes, m is the set of links, and the function F is a $n \times n$ mapping which defines the structure of the network. An entry of the $n \times n$ matrix is a zero in the row (or column) x and column (or row) y if the node x is not connected to node y. If connected, the entry is a 1. As introduced before, the $n \times n$ matrix is known as the Adjacency Matrix associated with the network graph G. The Adjacency Matrix is a complete representation of the network at the topological level. The actual performance of the network depends on the specifications associated with each of the transmission lines and the traffic on the network. The properties of a network and its constituent elements can be characterized at a high level by several parameters as follows.

A node in a network graph is characterized by its nodal degree which is the number of links the node has that connect it to the graph. A higher degree node has a larger number of links making it harder to disconnect the node from the network unless all the links which connect the node to other nodes are compromised.

In a connected graph, any pair nodes in the network is connected by one or more links. Generally speaking, there will be a number of ways in which traffic can traverse between a pair of nodes. The shortest path (meaning the least number of links) between a given pair of nodes is called the path length. The average path length of a network is the average of all shortest paths of the network. In a network of n nodes, there will be n(n–1)/2 possible shortest paths.

The expression n(n–1)/2 is also the maximum number of links a network of n nodes can have. Such a network is called the fully interconnected network. In a fully interconnected network, the path length between any pair of nodes is 1 and the average path length of the network is also 1.

The number of links in a network determines if the network is dense or sparse. The maximum number of links possible in a network of n nodes n(n–1)/2. The density of the network is defined as a fraction of the actual number of links to the maximum number of links. For example, if a network has n nodes and m links, the density ρ of the network is given as,

$$\rho = 2m / (n(n - 1)) \tag{3.2}$$

Figure 3.5 shows the graphical representation of a five-node network. The nodal degree for each of the nodes, the path length between pairs of nodes, and the average path length of the network are specified in Table 3.1.

Fig. 3.5 Graphical representation of a network

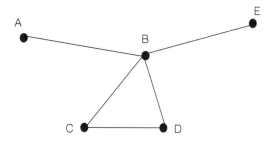

Table 3.1 Path lengths and average path length of a network

	A	B	C	D	E	Average path length
A	–	1	2	2	2	1.5
B	1	–	1	1	1	
C	2	1	–	1	2	
D	2	1	1	–	2	
E	2	1	2	2	–	

3.6 Degree Distribution and Clustering Property of Networks

We have defined the Degree of a Node in a network as the number of links that connect the node to the rest of the network. The distribution of the degrees of nodes in a network is an important parameter that can relate to the tolerance of the network against failures and attacks. The degree distribution of a network is one of its most basic properties. Mathematically, the degree k_i of the i_{th} node of a network is defined as,

$$k_i = \sum_{j=1}^{n} A_{ij} \tag{3.3}$$

where A_{ij} is the adjacency matrix associated with a network of n nodes.

If there are m links or edges in the network, then, $2m = \sum_{i=1}^{n} k_i$, or $2m = \sum_{i,j} A_{ij}$. This is because each link connects two nodes.

It also follows that the mean degree c of a network is given as,

$$c = \frac{1}{n} \sum_{i=1}^{n} k_i \tag{3.4}$$

or, from above,

$$c = \frac{2m}{n} \tag{3.5}$$

A network can be dense or sparse. If a network of n nodes has m links, its density ρ is defined as in Eq. (3.2) as $\rho = \frac{2m}{n(n-1)}$ in last section. This equation can be simplified as $\rho = \frac{c}{n-1}$, or, for relatively large n,

$$\rho = \frac{c}{n} \tag{3.6}$$

Networks can, of course, be more densely connected in specific parts of the network or around specific nodes.

The Cluster Coefficient of a node typifies how densely its adjacent nodes are connected to each other. Adjacent Nodes of a specific node are the nodes that are connected to the specific node by a single link.

Mathematically, the Cluster Coefficient of a node i is given as:

$$C_i = \frac{2E_i}{k_i(k_i - 1)} \tag{3.7}$$

where k_i is the number of nodes adjacent to node i and E_i is the actual number of links connecting these nodes to each other.

The cluster coefficient of a node i is an indicator of how densely the nodes adjacent to i are connected to each other.

3.7 Betweenness and Closeness of Network

Betweenness and Closeness of a node in a network are measures of importance of the node. Specifically, Betweenness of a node in a network is the number of *all* paths from all nodes to all other nodes that must pass through node. Closeness of a node is the number of all shortest paths from all nodes to all other nodes that must pass through the node. Clearly, not all nodes in a network are equally powerful in playing the roles of an intermediary. Betweenness and Closeness of a node play a vital role in defining its importance in preserving the integrity of the network. As an example, should a node be connected by a single link to the network, its importance to the network is limited to disconnecting its own communication with the network. On the other hand, if a node is more centrally located and has a nodal degree much above the average degree of nodes of the network, its malfunction will increase the path lengths of several node-to-node communication.

3.8 Diameter, Radius of a Node, and Center of a Graph

The longest path between any two nodes is defined as the Diameter of the network. The diameter of a line network with n nodes is, for example, equal to $n - 1$. The diameter of a star network is equal to 2. The diameter is an indicator of the worst case performance of the network since the two nodes that define the diameter of the network are farthest apart in terms of the number of links between them.

The Radius of a node in a network is the longest path length from the node to any other node. The largest radius in a network is thus its diameter.

The center of a network is the node with the smallest radius.

3.9 Entropy of a Network

As in many other scientific disciplines, e.g., thermodynamics or communication, entropy of a network is an indicator of randomness of the network. A mathematical descriptor of randomness in the graphical representation of a network measures the manner in which the nodal degrees of a network differ from each other. For example, when all the nodal degrees in a network are identical, such as in a ring network (See Fig. 3.2), or a fully interconnected network shown in Fig. 3.6, the entropy of the network is zero.

In telecommunications, if information is being communicated through a sequence of n symbols and the probability of occurrence of the symbol i is p_i, the entropy per symbol is given as:

$$E = -\sum_{i=1}^{n} p_i \log_2 p_i \tag{3.8}$$

Following a similar approach, the measure of randomness of a graph G is shown as:

Fig. 3.6 A five-node fully interconnected network

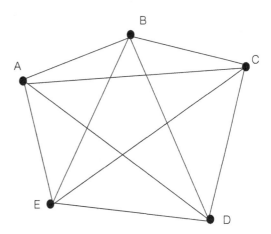

$$I(G) = - \sum_{i=1}^{max_{deg}} h_i \log_2 h_i \tag{3.9}$$

where h_i is the fraction of nodes with degree i.

3.10   Evolution of Networks

Design of legacy networks is based on the assumed need of users including their physical location and anticipated volume of the traffic they would generate or receive, the expected performance and, of course, the anticipated revenue that must provide an acceptable rate of return to the business. Network design uses results from graph theory, telecommunication traffic characteristics, and the cost of transmission facilities and switching nodes among others.

Contemporary networks, on the other hand, appear to form somewhat randomly. This is especially the case for social networks, citation networks, the World Wide Web, and the Internet which provides the base structure for information networks. With increasing mobility, the nodes and transmission facilities are no longer static but continuously evolve in a fashion that appears random. How can we characterize such networks in a way that can lead to a better understanding of their behavior? For example, is a particular class of networks more vulnerable to failure? How can we make such networks more tolerant against failure of links or nodes? Alternatively, how can we exploit the vulnerability of a network to break it totally or partially? This knowledge would be useful in the case of a biological network through which a communicable disease spreads and therefore it must be disabled or destroyed.

There was little understanding of the evolution or characterization of contemporary networks until recently. It is easy to classify any phenomenon as a random phenomenon if we cannot predict its occurrence with a level of precision we are comfortable with. All networks that defied characterization were, initially, classified as random networks.

The random network theory to characterize contemporary networks was first proposed in 1959 by Erdos and Renyi [2]. It dominated scientific thinking until just about 20 years ago. It offered the basis on which networks of roads, electric grids, or communication would evolve. random networks also included social networks, especially with the ability of an individual to connect to another irrespective of distance (or even language) in a medium of their mutual choice. The following section introduces some characteristics of random networks.

3.11 Models of Contemporary Networks: Random Networks

Models of contemporary networks are based on real-world systems. The World Wide Web or networks of friendship (each node representing a person and friendship represented by links) are examples of such networks. How do such networks evolve and what are their characteristics? We do not know for sure, but mathematicians or sociologists when they discover a phenomenon that cannot be tied down with precision in a statistical sense often name it a random phenomenon. Our knowledge of the properties of real networks has grown a lot over the past couple of decades; however, random networks still form a strong basis against which we can scale other networks. We discuss random networks in the following sub-sections.

3.11.1 Random Networks

Random networks constitute a class of networks where the deployed nodes are randomly connected to each other. One way to think of these networks is to think of a given number of links that are thrown randomly on the set of pair of nodes in a graph. Although real networks are not random, random networks offer a baseline against which a comparative evaluation of real networks can be carried out. Random networks thus form an important base case.

Random networks have been extensively studied by mathematicians. Erdos and Renyi first predicted and proved that the characteristics of random networks are almost always predictable with a high level of accuracy. Since, in a random network, the node pairs to be connected are chosen at random from the set of all possible node pairs, some of the nodes will have larger number of links than others. In other words, the degrees of the nodes will vary across the network.

Figure 3.7 shows the examples of two random networks, each with 20 nodes. Figure 3.7a shows a random network with 20 nodes and 19 links. Since the maximum number of links in a random network of 20 nodes can be $20(20 - 1)/2$ or 190 links, the probability p that a given pair of nodes is connected is 19/190 or 0.1. If we increase the number of links to 30 shown in Fig. 3.7b, the probability of a node pair being connected is close to 0.16. We do observe that even with a relatively small increase in the number of links, there is a sizeable difference between the connectivity properties of these two networks. We see that several nodes in Fig. 3.7a are disconnected while all the nodes in Fig. 3.7b are part of the network.

Even though Fig. 3.7 is just an example, and examples cannot be generalized in a mathematical sense, the observation that, even with a modest number of links (against the maximum possible number of links to fully interconnect a network) the nodes in a network are almost always connected, holds true.

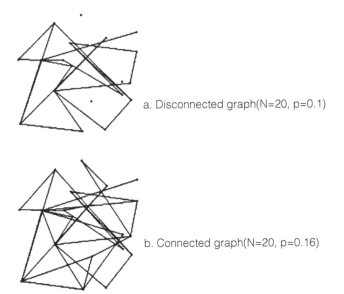

a. Disconnected graph(N=20, p=0.1)

b. Connected graph(N=20, p=0.16)

When p reaches a threshold, almost all graphs are connected.

Fig. 3.7 Two random networks

3.11.2 *Properties of Random Networks*

The graph $G(n, m)$ of a random network with n nodes and m links has an average degree $<K>$ given as $<K> = 2m/n$. This results holds for any network, as shown in Eq. (3.5).

If the number of edges in the network graph G is not known and only the probability p of having an edge between a pair of nodes is known, the graph is denoted as $G(n, p)$ where p is the probability of having a link between any pair of nodes. This means that the number of graphs with exactly n nodes and m edges is simply equal to the number of ways in which m edges can be picked from the $n(n-1)/2$ possible node pairs. The probability of having a graph with m links can thus be given by,

$$p(m) = \binom{\frac{n(n-1)}{2}}{m} p^m (1 - p)^{\frac{n(n-1)}{2} - m} \tag{3.10}$$

The mean value of m can now be written as, $\sum_{m=0}^{\frac{n(n-1)}{2}} p(m)$, which is equal to $\frac{n(n-1)p}{2}$.

The mean degree of a node can now be written as $<K> = (n - 1)p$.

The distribution of degrees in a random network can now be given as,

$$p_k = \binom{n-1}{k} p^k (1-p)^{n-1-k} \tag{3.11}$$

For large n, p_k can be simplified as,

$$p_k = \frac{(pn)^k e^{-pn}}{k!} \tag{3.12}$$

The average path length l_{rand}, i.e., the mean number of links needed to traverse all possible pairs of nodes can be shown to scale as,

$$l_{rand} \propto \frac{\ln(n)}{\ln(<K>)} \tag{3.13}$$

In other words, the average path length l_{rand} scales with the logarithm of n.

Equation (3.13) shows that the average path length in a random network is proportional to the logarithm of the number of nodes. This is an important result since it shows that, relatively speaking, larger networks do not have correspondingly large number of links between source-destination pairs.

The clustering, closeness, radius of a random network are as follows:

$$C_{rand} = p = \frac{<K>}{n-1} \tag{3.14}$$

Equation (3.14) shows that the clustering coefficient C_{rand} of random networks is low because the probability of having a connection between any two neighbors of a node is only p.

Closeness in random network can be shown to be [3],

$$closeness = O((1-density)\lambda^r) \tag{3.15}$$

where r is the diameter of the network and the sparseness of the network is $(1-density)$ or $(1-p)$, and $\lambda = np = \frac{2m}{n-1}$ through Eq. (3.2).

The radius of a random network can be shown to be [3],

$$r = \frac{A \log(n)}{\log(\lambda C) + D} \tag{3.16}$$

where A, C, D are the approximation parameters. The logarithm is to the base 2.

The entropy of a hypothetical random network is shown in Fig. 3.8. It can be seen from this figure that when the network has zero density, that is, when the network has no connectivity, it has zero entropy since all the nodes have identical degree, namely, zero. At the other extreme, when the network is fully interconnected, all the nodes have all identical degree, resulting in zero entropy for the network. Random topology leads to high entropy since the degree distribution of nodes is random.

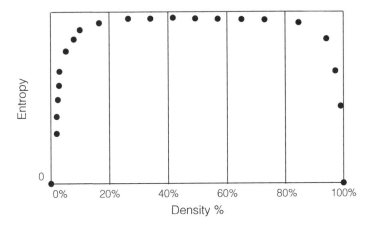

Fig. 3.8 Random network entropy vs. density

3.12 Summary

This chapter has characterized communication networks from a graph theoretic standpoint. It has presented the topology of several legacy networks. It has introduced random networks and characterized them using results from probability and statistics. Although real networks are seldom random, random networks form an important base against which the behavior of other networks is scaled.

The biggest shortcoming of random networks is that in such networks no distinction is made between any two nodes. Real networks display distinguishing characteristics of nodes and their preference to connect to nodes with distinguishing characteristics.

As an empirical observation, real networks have much higher clustering coefficients than random networks for an identical number of nodes and links. In other words, the nodes in the vicinity of a given node are also much more likely to be connected to each other. As, for example, in social networks, friends of a particular friend are much more likely to be friends of each other. Or, considering the example of a community of residents, neighbors of a resident are also neighbors of each other. The following chapter considers two prominent examples of real networks—the small world network and the scale-free network.

Problems

3.1 A fully interconnected telecommunication network has 36 edges. How many nodes does it have? What is the entropy of the network?

3.2 The entropy of a random network drops as the number of links approaches the maximum. Explain in plain words why this would likely be the case?

3.3 Explain qualitatively why the entropy of a random network is, roughly, the highest when the density is about 50% and is lowest at the ends where the density is either zero or 100%.

References

1. A. Barabasi, *Linked* (Perseus Publishing, Cambridge, 2002)
2. P. Erdos, A. Renyi, On the evolution of random graphs. Pub. Math. Inst. Hungar Acad. Sci. **5**, 17–61 (1960)
3. T. G. Lewis, *Network Science* (Wiley, 2009)

Chapter 4
Small World and Scale-Free Networks

4.1 Small World Networks

The foundation of small world networks was set by an experiment conducted by Harvard Professor Stanley Milgram in the 1960s [1]. This indeed was a groundbreaking study. Fundamentally, Milgram's experiment was to determine how many acquaintances it would take to establish a connection on average between two randomly selected people in the USA. The experiment carried out by Milgram resulted in what is now famously known as 6° of separation between any two arbitrarily chosen individuals. Stated in other words, it means that six hand-offs will likely result in making communication between unknown parties in the USA.

Most real networks are not random. Real networks are characterized by two attributes: They grow; in other words, they are not static, and, furthermore, they attach to other members not randomly, but preferentially. The preferential attachment of real networks makes it difficult to analyze them mathematically unlike the case with random networks.

Two characteristics of small world networks stand out: They have high clustering coefficient and a short average path length compared to their random network counterpart with the same number of nodes and links.

Clustering is a ubiquitous property of all real networks. In 1998, Watts and Strogatz [2] proposed a model of small world networks that has been extensively analyzed. The model includes both randomness and clustering. It starts with a ring lattice of n nodes. Each node is connected to its nearest $2k$ close neighbors, k neighbors on each side. A total of pkn additional links are then added on to the network randomly so that self-connections and duplicates are avoided. Figure 4.1 shows the network starting from $k = 1$ to $k = 3$. Only five random edges have been added in each case. The value of p, accordingly, changes from 0.25 for Fig. 4.1a to 0.125 for Fig. 4.1b and 0.083 for Fig. 4.1c.

The surprising result associated with Fig. 4.1 is that even with a small number of additional links, the average path length is drastically reduced. The reason behind

© Springer Nature Switzerland AG 2020
P. Verma, F. Zhang, *The Economics of Telecommunication Services*, Textbooks in Telecommunication Engineering, https://doi.org/10.1007/978-3-030-33865-7_4

Fig. 4.1 Networks with
added edges

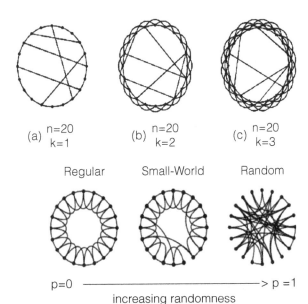

(a) $\begin{matrix} n=20 \\ k=1 \end{matrix}$ (b) $\begin{matrix} n=20 \\ k=2 \end{matrix}$ (c) $\begin{matrix} n=20 \\ k=3 \end{matrix}$

Fig. 4.2 Modified Watts and
Strogatz networks with added
edges

Regular   Small-World   Random

p=0 ————————————————————> p =1
increasing randomness

this is that even a few long edges that are added provide shortcuts that connect relatively distant nodes. For small values of p, the average path length in the small world network scales linearly with the number of nodes while, for large values of p, the scaling is logarithmic.

In 1999, Newman and Watts [3] modified the Watts and Strogatz model. The difference between the two models is that in the Newman and Watts model, edges are removed and an equivalent number of edges added. In the Watts and Strogatz model, no edges were removed and a total of pkn links were added. The modified WS model is shown in Fig. 4.2. As can be inferred, as the number of randomly placed links increases, randomness of the Newman and Watts model increases and its characteristics come closer to a random network.

Both the random and the small world networks exclude one property of real networks. This property relates to the propensity of large-degree nodes to also connect to each other. As an example, social networks which are networks of real people, influential people are connected to other influential people with higher probability. This feature is captured in scale-free networks which is also known as the network of hubs and connectors.

The small world networks are based on networks where each node to start with has the same degree with some additional random edges superimposed. Or, alternatively, some edges are removed randomly and an equivalent number of edges added, randomly. The nodal degree of small world networks thus changes within a small range. As such, the entropy of small world networks compared to random networks with identical number of nodes and links is smaller.

The average path length of small world networks is, however, smaller than that of a corresponding random network. The reason is that a few long edges that are added to the small world network provide shortcuts that results in reducing the average path length considerably.

In the Watts and Strogatz model, for small values of p, the average path length scales linearly with the size n. For large values of p, the scaling is logarithmic.

Neither random network nor small world networks support the concept of hubs and connectors that attract and connect to a large number of other nodes. The existence of hubs is common in social networks where there are people who have a large number of friends and, therefore, wield a higher level of influence. The existence of hubs and connectors, typical in naturally evolving networks, is a characteristic of scale-free networks addressed in the following section.

4.2 Scale-Free Networks

Scale-free networks follow a power law distribution of the degree of a node. Mathematically, the probability $p(k)$ that a node has a degree k is proportional to $k^{-\alpha}$, where α is a constant; i.e.,

$$p(k) \propto k^{-\alpha} \tag{4.1}$$

In 1965, Price [4] discovered that the degree distribution of citation networks (i.e., the number of citations a paper receives) follows a power law. More recent research in the 1990s has shown that a number of real networks follow the power law degree distribution. This includes the Internet and the WWW.

In the 1990s, a number of researchers Watts, Strogatz, Barabasi, and others provided a strong stimulus in creating a new discipline called Network Science. The Barbasi–Albert model proposed in 1999 is captured here. The model is based on the following algorithm.

Growth Starting with a small number of n_0 nodes, each time a new node is added, it forms $n(\leq n_0)$ links with other n different nodes already present in the network.

Connectivity The probability $p(k_i)$ that a new node will connect to node i depends on the degree k_i of node i, such that,

$$p(k_i) = \frac{k_i}{\sum_i k_i}.$$

A scale-free network is shown in Fig. 4.3. The evolution of the first few steps, carried out through a simulation, is captured in Fig. 4.4.

Scale-free networks include a few (but finite) number of nodes which serve as connectors or hubs. It has a short average path length. It can be shown that for a large n, the average path length approaches,

Fig. 4.3 Scale-free networks

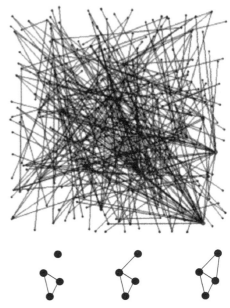

Fig. 4.4 Scale-free network
simulation

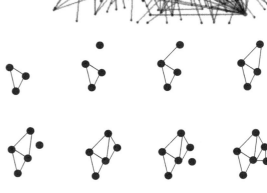

$$\frac{\log(n)}{\log(\log(n))}.$$

The scale-free network has low clustering coefficient compared to the small world network, but higher than the clustering coefficient of a random network.

The average closeness of the scale-free network, as its density increases is, generally, as follows. First, with increasing density, the average closeness increases to a peak value. However, beyond a peak at 50% of density, the average closeness declines. This is because more and more nodes have direct connection with the dominant hub. Compared to random and small world networks, the average closeness of the scale-free network is much lower.

A compelling characteristic of scale-free networks is that a significant fraction of nodes can be removed randomly without the network loosing connectivity. On the other hand, disabling just a few critical nodes can result in a disconnected network especially for the smaller nodes.

Table 4.1 presents a comparison of the properties of random, small world, and scale-free networks for a network of 200 nodes. For the scale-free network, $n_0 = 5$, $n = 3$. For small world network, $k = 2, p = 0.5$.

Table 4.1 Comparison of some properties of networks

Property	ER	BA	WS
Hub degree	14	37	10
Average degree	6	5.93	5.99
Distribution	Poisson	Power	Poisson-like
Average path length	3.13	2.88	3.44
Diameter	6	5	6
Cluster coefficient	0.023	0.091	0.25
Entropy	3.3	2.9	2.3

Scale-free networks characterize several networks that evolve naturally. In particular, the Internet has been shown to follow the power law distribution. The exponent alpha of the power law is known to vary between 2 and 3 for most power law based networks.

One notable characteristic of power law networks is its long tail accompanied by a finite number of nodes with large degrees. In other words, a small but finite number of hubs connect to a large number of nodes. This property reduces the average path length considerably since the hubs also tend to be connected to each other.

It is easy to observe that Scale-free networks will be very vulnerable to targeted attacks that will destroy one or more network hubs, since such an attack can also disconnect several smaller nodes from the network. On the other hand, a random attack on nodes will likely destroy a node with connectivity to just a few other nodes resulting in containing the impact on the larger network.

4.3 Summary

This chapter has discussed and characterized two classes of networks: the small world network and the scale-free network. Unlike random networks, these two classes of networks characterize several real-world networks and natural phenomena. We have also observed that scale-free networks are susceptible to targeted attacks but are relatively immune to random attacks.

Problems

4.1 Name the three generic classifications of complex networks. Name one defining characteristic of each network.

4.2 Differentiate among random, small world, and scale-free networks. How do their key characteristics differ?

4.3 The scale-free network is considered more vulnerable to targeted attacks. Why?

4.4 Random attacks are carried out on each of the three generic classifications of networks. Which network will have the least vulnerability to random attacks? Why?

References

1. S. Milgram, The Small World Problem. Psychol. Today **2**, 60–67 (1967)
2. D.J. Watts, S.H. Strogatz, Collective dynamics of "small world" networks. Nature **393**, 440–442 (1998)
3. M.E.J. Newman, D.J. Watts, Scaling and percolation in the small world network model. Phys. Rev. E **60**(6), 7332–7342 (1999)
4. D.J.D.S. Price, Networks of scientific papers. Science **149**, 510–515 (1965)

Chapter 5
Characterization of Telecommunication Traffic

5.1 Common User Networks

Telecommunication services are almost always offered through common user networks. A user demands services when needed and expects to be provided the service requested at an acceptable level of performance and price. Common user networks are not expected to meet performance objectives on a guaranteed basis. The customer, however, expects that the common user network will meet performance objectives on a statistical basis. A common user network that will guarantee connectivity among all possible source destination pairs at a guaranteed level of performance will be cost prohibitive for all practical purposes.

Nodes of the network constitute switching or routing points of the network, and are responsible for directing traffic on a link toward destination. The links of the network offer pathways with predefined rates of data transfer between the nodes they connect. The unavailability of processing resources at a node or congestion at an outgoing transmission line can result in communication failure or unacceptable delay.

5.2 Circuit and Packet Switching

Circuit and Packet Switching form the two main categories of switching bit streams or packets at a node. In circuit switching, the end-to-end path must be established before transferring data from the source. Once established, the path will remain open until all the data has been transferred from the source. If an end-to-end path is not established, no data is transferred and another attempt must be made to reestablish the path. This discipline to transfer data is called blocked and lost paradigm and is experienced by any telephone user when the called entity or any intermediate switch or link (also called a trunk in telephone parlance) is not

© Springer Nature Switzerland AG 2020
P. Verma, F. Zhang, *The Economics of Telecommunication Services*, Textbooks in Telecommunication Engineering, https://doi.org/10.1007/978-3-030-33865-7_5

available to complete the end-to-end path. Circuit switching was originally used for public switched networks for voice traffic. High Speed optical networks such as Synchronous Optical Networks (SONET) using Dense Wave Division Multiplexing (DWDM) techniques use circuit switching as the core transmission resource.

In Packet Switching, information bits associated with a message to be transferred are divided and put into packets sequentially. The packets, as entities in their own right, travel from node to node to the destination while carrying their respective addresses. At the destination, the message is assembled and delivered to the receiving party.

In circuit switching, once established, the transmission resource is guaranteed to remain available until the end user has completed the transmission. The overhead in circuit switching is very little because it is only needed during the process of establishing or tearing down the end-to-end path. In packet switching, each packet, as it travels toward its destination, carries information concerning its ultimate destination. This makes the overhead a significant fraction of the data to be transferred. In packet switching, unlike circuit switching, connections are not blocked when a path from one node to another is not available. The packet is simply stored until the path becomes available. Packed switching thus forms the blocked and delayed paradigm for transmission.

5.3   Performance Parameters of Networks

There are four categories of performance parameters from the users' perspective. These are:

- Delay related performance parameters or delay parameters
- Throughput related performance parameters or throughput parameters
- Integrity related performance parameters or accuracy parameters, and
- Availability related performance parameters, or availability parameters

Delay parameters are a class of parameters that reflect the duration of time between the availability of information at the source and its delivery at the destination. Throughput parameters quantify the volume of data a user can transfer per unit of time. In designating throughput, the absolute delay between the transmission and reception of information is relatively unimportant. It might appear at first that delay and throughput are inversely related. However, this is not the case. As an example, a satellite channel has a large static delay associated with it. However, it can provide a very high throughput for the transfer of one-way information, or two-way information with the use of an appropriate protocol.

Integrity related parameters or accuracy parameters capture the error the transmitted information has suffered in the transmission process. Bit error rate (number of bits in error divided by the total number of bits transmitted) is one parameter that is an indicator of the integrity of the transmission channel.

In a circuit switched network, the delay incurred in the end-to-end transmission, for all practical purposes, is constant. For packet switched networks, the end-to-end delay is composed of a number of elements. These are:

- Packetization delay: The delay incurred in the formation of a packet because an incoming bit must be buffered until the sending entity has received enough bits that would form a packet. Once a packet is formed, it is transmitted at the speed of the transmission link.
- Transmission delay: The time it takes for a complete packet to empty on to the transmission link. It is equal to the length of the packet in bits divided by the speed of the transmission line in bits per second.
- Propagation delay: Propagation delay is the signal propagation time and is a function of the medium of transmission and the distance the packet must travel. In any medium, the propagation time is the distance divided by the speed of electromagnetic waves in that medium.
- Processing delay: Any operation on a packet at the point of origin or during transit will entail a delay. For example, one component of processing delay is the time required for a switching device to process packet headers in order to make routing decisions.
- Queuing delay: Packets in a packet switched networks suffer delay at each transit node if they are buffered because the outgoing link is backed up. Queuing delays are incurred when packets arrive faster than the speed at which the switching device can forward them toward destination. Queuing delay is a function of the line speed and traffic intensity. Queuing delays can be reduced by increasing the capacity of transmission links.
- Depacketization delay: Every arriving packet at the receive end is buffered until its last bit has arrived. The delay incurred in the process is the depacketization delay.

A communication system is termed unavailable when it is not able to function as designed or when its performance is below a threshold that the user considers acceptable. Recovery from an unavailable condition might be automatic or manual such as when a working part might have to be replaced manually. Availability is generally expressed as a fraction or percentage of the time the system is unavailable divided by the sum of the time it is available plus the time it is unavailable.

In packet switched networks, end-to-end packet transfer time is not constant and is affected by varying queuing delays it suffers at the intermediate nodes. This introduces jitter in the system. For real-time applications, jitter is very annoying. For non-real-time applications, jitter can be accommodated by introducing buffers which fill in at variable rates by incoming traffic but are forwarded at a constant rate to the receive end point.

5.4   Characterization of Traffic

The manner in which traffic originates, specifically, the points in time at which service requests are made, and the distribution of the volume of originating traffic are the two parameters that characterize traffic.

The classical model of traffic origination is based on Markovian stochastic models, in particular, the Poisson process. The Poisson process is a memoryless process meaning that traffic originates at random points in time.

The memoryless model also applies as far as the volume of traffic, usually designated by its length in bits, is concerned. This implies that most of the traffic length is close to the mean and the probability of long lengths of traffic (designated in bits) sharply reduces. One advantage of the Poisson model is that it is mathematically tractable and fairly reflects the real-world traffic demand.

The Poissonian model has come into increasing attack with the discovery in the early 1990s of the self-similar nature of telecommunication traffic. Self-similarity implies that there is correlation between the arrival times of traffic at the point of origin of traffic. Furthermore, unlike the Poisson model which suggests that the probability of large bursts of traffic decreases exponentially with the size of traffic, self-similar traffic can have large bursts with distinctly higher probability compared to the Poisson model.

Despite the difference between the analytical approaches to deal with random or self-similar traffic, this book largely follows the Poisson model. The Poisson model is mathematically tractable and offers insight into the behavior of telecommunication networks. Generally speaking, resource requirements for telecommunication networks based on the Poisson model, when compared to self-similar models, offer an optimistic estimate.

Traditional characterization of circuit switched traffic has been in terms of Erlangs which reflects the proportion of time a transmission link is busy. The Erlangian model of traffic characterization was useful when demands on a transmission link were slotted as opposed to being aggregated. Contemporary transmission systems are largely aggregated up to the capacity of a wavelength in a dense wavelength division multiplexed fiber-optic transmission system. If a wavelength carrying traffic at 10 Mbps is busy 50% of the time on average, the traffic is 0.5 Erlang.

Digital traffic is characterized in bits per second. The volume of traffic relative to the carrying capacity of the transmission line is designated as the intensity of traffic. As an example if packets are arriving on average at the rate of 50 packets a second and the average length of a packet is 100 kb then a 10 Mbps transmission link will experience an intensity of (50 * 100, 000)/(10 * 1000, 000) or 0.5.

5.5 Selection of Transmission Medium

Different transmission media have different error performance characteristics. A service provider must choose a medium that offers the best trade-off between cost of providing the service and the revenue the provider would generate.

In the following illustrative example, we compare two transmission systems: System$_1$ and System$_2$, with error rates of p_1 and p_2, respectively, and spectral efficiencies of α_1 and α_2. Spectral efficiency is expressed in bits per second per Hz and typifies the ability of the system to forward digital information in unit analog bandwidth.

The two alternatives in the choice of transmission media are depicted in Table 5.1. For the sake of simplicity, we assume that transmission delay is not important and no error correction techniques are deployed. Furthermore, we also assume that the cost of transmission is proportional to the bandwidth in Hz; and for every bit correctly transferred, the provider generates one unit of revenue. On the other hand, for every bit in error in the transmission process, the provider is required to pay back 10 units of revenue. The provider needs to make a judicious choice between the two transmission media.

Let the payload (number of bits to be transferred) be N bits. Assume the service provider buys bandwidth (in Hz) for either System$_1$ or System$_2$ at a cost C. Then,

System$_1$ transfers $N(1 - p_1)\alpha_1$ bits correctly and $Np_1\alpha_1$ bits incorrectly. The net revenue generated by System$_1$ is, therefore,

- $NR_1 = N[(1 - p_1)\alpha_1 - 10p_1\alpha_1] - C$, or
- $NR_1 = N[\alpha_1 - 11p_1\alpha_1] - C$.

Similarly, the net revenue generated by System$_2$ is,

- $NR_2 = N[(1 - p_2)\alpha_2 - 10p_2\alpha_2] - C$, or
- $NR_2 = N[\alpha_2 - 11p_2\alpha_2] - C$.

System$_1$ is therefore preferable to System$_2$ if,

- $NR_1 > NR_2$, or if,
- $\alpha_1 - 11p_1\alpha_1 > \alpha_2 - 11p_2\alpha_2$, or if,

$$\frac{\alpha_1}{\alpha_2} > \frac{1 - 11p_2}{1 - 11p_1} \tag{5.1}$$

Equation (5.1) shows the implication of a transmission medium's spectral efficiency as well as its error rate in its selection process. As a somewhat trivial result from Eq. (5.1), one can note that if the error rates in the two systems are identical,

Table 5.1 Characterization of two transmission systems

System	Error rate	Transmission efficiency (bits/s/Hz)
System$_1$	p_1	α_1
System$_2$	p_2	α_2

the one with a higher spectral efficiency is desirable. A similar approach can be adopted if the cost per unit of bandwidth of $System_1$ and $System_2$ are different, which is likely. Interactive systems, such as human speech, are much less tolerant of absolute delay, or jitter. On the other hand, applications like email or bulk data transfer can easily tolerate them.

5.6   Interdependence of Performance Parameters

The performance parameters of networks are closely inter-related. The end-to-end error performance requirement is, most generally, a function of the specific application. Some applications such as speech can tolerate a higher error threshold because of the redundancy inherent in human conversations. In others, such as in financial transactions, the loss or mutilation of even a single digit might have disastrous consequences.

Correction of errors in the transmission process can be accomplished by either forward error correction or backward error correction. Either of these error correction techniques results in lower throughput and enhanced delay. In forward error correction, an appropriate number of redundant bits are introduced along with the user data in such a way that the function of detecting data bits in error as well as correcting those bits are both accomplished in a single step. On the other hand, backward error correcting scheme first detects errors such as by using the cyclic redundancy code [1] in a packet-based transmission system. The packet in error is then retransmitted from the sending end to the receiving end. The process can be repeated if the retransmitted packet is detected to have an error. The number of overhead bits in the backward error correcting scheme is, usually, much smaller than that in the forward error correction scheme.

5.7   Forward Error Correction Scheme

Unlike the backward error correction scheme, forward error correction techniques do not need a reverse communication channel to intimate a transmission failure to the sending node. Furthermore, these techniques are especially useful when bits in error are relatively independent. The development of forward error correction schemes is an extensive field of mathematics [1, 2] and is not discussed here with any rigor. We illustrate the technique by a simple example as follows.

Consider a user who needs to transmit a sequence of 16 bits 1001101001001111. We first organize the bits into a 4×4 matrix as shown in Table 5.2.

We now generate another matrix shown in Table 5.3 by adding an additional bit in each column and each row such that the number of 1's in each row and column is even.

Table 5.2 Transmission
Matrix1

1	1	0	1
0	0	1	1
0	1	0	1
1	0	0	1

Table 5.3 Transmission
Matrix2

	0	0	1	0
1	1	1	0	1
0	0	0	1	1
0	0	1	0	1
0	1	0	0	1

While transmitting, the sending entity transmits a total of 24 bits as shown in the matrix above. The receiving entity reconstructs the matrix transmitted, clearly identifying the data bits and the overhead bits. If any single bit of data transmitted is corrupted in the process of transmission, one single row and one single column will show violation of parity. This will clearly identify the bit in error which can be reversed and the correct bit recovered.

If, on the other hand, a single bit error occurs in the overhead bits, there will be an odd number of 1's in some column or some row, but not in both as was the case when a data bit was in error. This will imply that the one of the check bits was in error which can then be discarded. If more than one check bits are in error, the process will not work reliably.

One can, of course, include a larger number of customer originated bits and keep the overhead bits limited to one for each row and for each column. For example, a 32×32 user data grid will have 1024 data bits with only 64 over head bits. This will result in the ratio of overhead bits to user data bits to only 6.25%. From a transmission efficiency standpoint, such a scheme will be more efficient. However, with a larger number of bits, the probability of having more than one bit corrupted in the transmission process will correspondingly increase. This will compromise the effectiveness of the scheme in producing correct data at the receiving end.

Most transmission systems are subject to high correlation among bits in error. In other words, if a bit is in error, the probability that the nearby bits are also in error is high. This results in error bits being bunched together. The forward error correction scheme is not effective in such a situation. The backward error correction scheme that corrects errors by retransmitting a packet in error as a whole is much more effective. Furthermore, the backward error correction scheme has a considerably less number of overhead bits. Correction of errors encountered in transmission in the backward error correction scheme is obviously at the cost of delay incurred due to retransmission of packets in error and the number of overhead bits associated with each packet. The following section offers a quantitative treatment on the delay and throughput associated with the backward error correcting scheme.

5.8 Backward Error Correction Scheme

In order to quantify the impact of backward error correction on delay and through-put, we assume that the bits in error are randomly distributed. Let,

- $n =$ number of data bits in a packet
- $s =$ speed of the transmission line in bits per second
- $r =$ number of overhead bits associated with each packet in order to detect a packet in error
- $p =$ bit error rate of the transmission medium (the bits in error are assumed to be independent)
- $d =$ transmission delay associated with the transmission medium in any one direction

Since the bit error rate is p, the probability that a bit is correctly received is $1 - p$. This gives, then, the probability of correct reception of all bits in a packet as $(1 - p)^{(n+r)}$.

The probability P that there is at least one bit in error in a packet is thus given as, $P = 1 - (1 - p)^{n+r}$. P is also the probability that a block will be retransmitted.

We also assume that there is negligible time associated with detecting a packet in error and that the retransmission of the packet in error begins immediately after the sending end is notified about the packet in error.

It is entirely possible that a retransmitted packet suffers an error again and has to be retransmitted one or more times. Such retransmissions will occur with exponentially reducing probabilities, however. Let the transmission delay in any direction be fixed and equal to d.

We can now derive the following results.

Case 1: No Transmission Error in Any Packet
- Probability $= 1 - P$
- Number of user data bits $= n$
- Total number of bits transmitted $= n + r$
- Total delay $= \frac{(n+r)}{s} + d$
- Relative throughput $= \frac{n}{(n+r)}$

Case 2: Correct Reception After a Single Transmission
- Probability $= P(1 - P)$
- Number of user data bits $= n$
- Total number of bits transmitted $= 2(n + r)$
- Total delay $= \frac{2(n+r)}{s} + 3d$
- Relative throughput $= \frac{n}{2(n+r)}$

Case 3: Correct Reception After Two Retransmissions
- Probability $= P^2(1 - P)$
- Number of user data bits $= n$
- Total number of bits transmitted $= 3(n + r)$
- Total delay $= \frac{3(n+r)}{s} + 5d$
- Relative throughput $= \frac{n}{3(n+r)}$

The sequence can be generalized so that for correct reception after m retransmissions, we have,

- Probability $= P^m(1 - P)$
- Number of user data bits $= n$
- Total number of bits transmitted $= (m + 1)(n + r)$
- Total delay $= \frac{(m+1)(n+r)}{s} + (2m + 1)d$
- Relative throughput $= \frac{n}{(m+1)(n+r)}$

The mean time taken to send a block can thus be given by the sum of the delays weighted by their probabilities, i.e.,

$$
\begin{aligned}
Mean\,Delay &= \sum_{m=0}^{\infty} P^m (1 - P) \left[\frac{(m + 1)(n + r)}{s} + (2m + 1)d \right] \\
&= \frac{1 - P}{s} \sum_{m=0}^{\infty} P^m \left[m(n + r + 2sd) + (n + r + sd) \right] \\
&= \frac{1 - P}{s} \left[\sum_{m=0}^{\infty} P^m \left[m(m + r + 2sd) \right] + \sum_{m=0}^{\infty} P^m (n + r + sd) \right] \\
&= \frac{1 - P}{s} \left[(n + r + 2sd) \sum_{m=0}^{\infty} P^m m + \frac{n + r + sd}{1 - P} \right] \\
&= \frac{1 - P}{s} \left[(n + r + 2sd)(P + 2P^2 + 3P^3 + \cdots) + \frac{n + r + sd}{1 - P} \right] \\
&= \frac{1 - P}{s} \left[P(n + r + 2sd)(1 + 2P + 3P^2 + \cdots) + \frac{n + r + sd}{1 - P} \right] \\
&= \frac{1 - P}{s} \left[P(n + r + 2sd) \frac{1}{(1 - P)^2} \frac{n + r + sd}{1 - P} \right] \\
&= \frac{P(n + r + 2sd)}{s(1 - P)} + \frac{n + r + sd}{s} \\
&= \frac{n + r + sd(1 + P)}{s(1 - P)}
\end{aligned}
$$

$$(5.2)$$

In the absence of any error and the use of any error correction scheme, the delay in transmitting the packet will be,

$$\frac{n + sd}{s} \tag{5.3}$$

The relative delay can be given as:

$$Relative\,Delay = \frac{n + r + sd(1 + P)}{(n + sd)(1 - P)} \tag{5.4}$$

It can be seen that the relative delay is always greater than 1; the asymptotic value of 1 is reaching when there are no overhead bits, that is, $r = 0$ and the block error rate approaches zero. The relative throughput can similarly be computed as:

$$
\begin{aligned}
Relative\,Throughput &= \sum_{m=0}^{\infty} P^m (1 - P) \frac{n}{(m + 1)(n + r)} \\
&= \frac{n(1 - P)}{n + r} \sum_{m=0}^{\infty} \frac{P^m}{m + 1} \\
&= \frac{n(1 - P)}{n + r} \left[1 + \frac{P}{2} + \frac{P^2}{3} + \frac{P^3}{4} + \cdots \right] \\
&= \frac{n(1 - P)}{(n + r)P} \left[P + \frac{P^2}{2} + \frac{P^3}{3} + \frac{P^4}{4} + \cdots \right] \\
&= \frac{n(1 - P)}{(n + r)P} \int_0^P (1 + P + P^2 + \cdots) dP \\
&= \frac{n(1 - P)}{(n + r)P} \int_0^P \frac{dP}{1 - P} \\
&= \frac{n(1 - P)}{(n + r)P} \ln \frac{1}{1 - P}
\end{aligned}
\tag{5.5}
$$

It can be shown that the relative throughput is always smaller than unity. Its value reaches 1 asymptotically when there are no overhead bits, that is, $r = 0$, and the block error rate P approaches zero.

Lecture Notes in Computer Science　11797

More information about this series at http://www.springer.com/series/7412

Kenji Suzuki · Mauricio Reyes ·
Tanveer Syeda-Mahmood et al. (Eds.)

Interpretability of Machine Intelligence in Medical Image Computing and Multimodal Learning for Clinical Decision Support

Second International Workshop, iMIMIC 2019
and 9th International Workshop, ML-CDS 2019
Held in Conjunction with MICCAI 2019
Shenzhen, China, October 17, 2019
Proceedings

 Springer

Editors
Kenji Suzuki
Tokyo Institute of Technology
Yokohama, Japan

Mauricio Reyes
University of Bern
Bern, Switzerland

Tanveer Syeda-Mahmood
IBM Research - Almaden
San Jose, CA, USA

Additional Workshop Editors *see next page*

ISSN 0302-9743   ISSN 1611-3349   (electronic)
Lecture Notes in Computer Science
ISBN 978-3-030-33849-7   ISBN 978-3-030-33850-3   (eBook)
https://doi.org/10.1007/978-3-030-33850-3

LNCS Sublibrary: SL6 – Image Processing, Computer Vision, Pattern Recognition, and Graphics

This Springer imprint is published by the registered company Springer Nature Switzerland AG
The registered company address is: Gewerbestrasse 11, 6330 Cham, Switzerland

Additional Workshop Editors

Satellite Events Chair

Kenji Suzuki
Tokyo Institute of Technology
Yokohama, Japan

Workshop Chairs

Hongen Liao
Tsinghua University
Beijing, China

Hayit Greenspan
Tel Aviv University
Tel Aviv, Israel

Challenge Chairs

Qian Wang
Shanghai Jiaotong University
Shanghai, China

Bram van Ginneken
Radboud University
Nijmegen, The Netherlands

Tutorial Chair

Luping Zhou
University of Sydney
Sydney, Australia

iMIMIC 2019 Editors

Mauricio Reyes
University of Bern
Bern, Switzerland

Ben Glocker
Imperial College London
London, UK

Ender Konukoglu
ETH Zurich
Zürich, Zürich, Germany

Roland Wiest ⓘ
University Hospital of Bern
Bern, Switzerland

ML-CDS 2019 Editors

Tanveer Syeda-Mahmood
IBM Research - Almaden
San Jose, CA, USA

Hayit Greenspan
Tel Aviv University
Ramat Aviv, Israel

Yaniv Gur
IBM Research - Almaden
San Jose, CA, USA

Anant Madabhushi
Case Western Reserve University
Cleveland, OH, USA

5.9 The Impact of Window Size on Throughput

With a finite value of delay associated with transmission of the packet in either direction, it is obvious that the transmission channel will be idle for periods of time if a packet-by-packet transmission was mandated and the next packet could not be sent by the sending end until a positive acknowledgment of the immediately preceding packet was received. The idle time on the channel would reduce throughput.

For this reason, most backward error correction systems in use allow a certain number of packets w, also known as the window size, to be outstanding before stopping transmission of the next packet. The size of the window is related to the delay d associated with the channel.

In order to derive a relationship between the related variables, let

- $n =$ number of bits per packet
- $s =$ channel speed (bits/s)
- $d =$ propagation delay (s)
- $w =$ window size

The transmission time required to send a packet is n/s. Because of the transmission delay associated with the channel, it will take an additional d seconds before the packet is received. The receiver will take an additional time t check the received frame and generate the acknowledgement signal which will arrive at the transmitting end d seconds later (assuming that the acknowledgement signal itself is sufficiently short).

The total delay associated with the receipt of an acknowledgement is, therefore,

$$\frac{n}{s} + 2d + t \tag{5.6}$$

In order to assume continuity of transmission, this time must be less than or equal to the transmission time associated with sending w packets, that is,

$$\frac{n}{s} + 2d + t \le w\frac{n}{s} \tag{5.7}$$

Equation (5.7) offers a relationship between the delay associated with transmission and window size.

5.10 Summary

This chapter has addressed the parameters in terms of which the user perceives the performance of a system. The architecture and the capacity of the network along with the characteristics of the traffic are the major determinants of performance perceived by the user. The chapter has developed relationships between and among the parameters that represent delay, throughput, integrity and availability of a system from a user's perspective.

Problems

5.1 A transmission system can transport n_1 messages of Class 1, n_2 messages of Class 2, n_3 messages of Class 3,..., n_n messages of class n per unit time. The probability of occurrence of class n message is p_n. Show that the number of messages carried by the system is the weighted harmonic mean of the number of messages from each class. The weighting is with respect to the probability of occurrence of the message of a class. In other words, if N is the total number of messages carried, then,

$$1/N = p_1/n_1 + p_2/n_2 + \ldots + p_n/n_n$$

Assume that each message of a particular class has the same number of bits.

5.2 Let p be the probability of occurrence of a bit in error. For a block of n bits, show that the probability of occurrence of one or more bits in error, P, is given by, $P = 1 - (1 - p)^n$ Assume that the bits in error are randomly distributed.

5.3 Let the bit error rate in a transmission system be 10^{-4}. Assuming transmission blocks of length 1000 bits, find the probability that a block will be in error.

5.4 Transmission performance is sometimes specified as %Error Free Seconds (%EFS). In other words, %EFS = Total error-free seconds in transmission/Total seconds of transmission. Show that when a block of transmission takes 1 s of transmission time, % EFS = %(1 − Block Error Rate).

References

1. G.C. Clark, Jr., J.B. Cain, *Error-Correction Coding for Digital Communications* (Plenum Press, New York, 1981). ISBN 0-306-40615-2
2. S.B. Wicker, *Error Control Systems for Digital Communication and Storage* (Prentice-Hall, Englewood Cliffs, 1995). ISBN 0-13-200809-2

Chapter 6
Bandwidth and Throughput of Networks: Circuit Switched Networks

6.1 Circuit and Packet Switched Traffic

The earlier chapters of the book have largely focused on the topological characterization of telecommunication networks and its impact on performance. Since
the prime objective of networking is conveyance of traffic, it is important to
understand the limits of a network in carrying a volume of traffic that meets a user's
performance expectation. Traditional telecommunication traffic has been described
in Erlangs. If traffic is being served by a pool of resources, each of which is identical,
the number of resources that are busy at a particular time is a measure of the intensity
of traffic at that time. This measure of the intensity of traffic is more suited for the
legacy PSTN where traffic between telecommunication central offices was served by
a bank of trunks, each trunk is capable of serving one customer. Using this measure,
the traffic intensity during a given period of time T is given as

$$E(T) = \frac{1}{T} \int_0^T n(t)dt \tag{6.1}$$

where $n(t)$ is the number of resources occupied at time t.

Equation (6.1) presents the traffic in Erlangs; the unit Erlang is in honor of the
Danish mathematician, A. K. Erlang (1878–1929), who first established a formal
way to measure telecommunication traffic. The traffic unit Erlang is simply a
positive number.

There are two terms that we will be using several times in this book: offered
traffic or incident traffic and carried traffic or served traffic. The meanings of the
terms are self-evident; it is important, however, to note that the customer pays only
for the carried or the served traffic.

For a circuit switched network, the offered traffic A can be written as

$$A = \lambda s \tag{6.2}$$

© Springer Nature Switzerland AG 2020
P. Verma, F. Zhang, *The Economics of Telecommunication Services*, Textbooks in
Telecommunication Engineering, https://doi.org/10.1007/978-3-030-33865-7_6

where λ is the average number of call attempts per unit time and s is the mean holding (or service) time.

Under ideal conditions, the offered traffic is also the carried traffic implying that every call that was attempted was also successfully completed. In other words, no call attempt was rejected because a needed resource in the network was not available. This is of course the ideal condition.

The contemporary measures of traffic, much more suitable for carrying high speed data in terms of bandwidth, are expressed as bits per second. The intensity of traffic, also called the utilization, in this case is expressed as a fraction that expresses the demand imposed on the system divided by the capacity of the system to meet that demand. Mathematically,

$$\rho = \frac{\lambda}{\mu C} \tag{6.3}$$

where ρ is the system utilization expressed as a dimensionless fraction, λ is the number of jobs (most generally expressed as packets) requesting service per unit of time, and $1/\mu$ is the average length of a packet expressed in bits. C is the capacity of the system, also referred to as the channel speed, in bits per second.

Numerical Examples

Erlangian Traffic Two telephone central offices are connected by 600 trunks in each direction, each of which can carry one voice channel. If the average number of calls from any direction is 100 calls per minute, each of 4 minutes' duration on average, what is the traffic intensity in either direction?

The offered traffic is 400 Erlangs. With 600 trunks, the traffic intensity is 400/600 or .667.

Data Traffic There are 500 data terminals, each generating traffic at the constant rate of 10 Mbps. If the traffic is concentrated into a single transmission line that has a capacity of 10 Gbps, what is the utilization of the line?

The offered traffic is 5 Gbps and the traffic intensity is 0.5.

The traffic intensity or utilization in the above examples is based on the notion of average over a period of time. Offered traffic would likely vary from instant to instant. Telecommunication systems are not designed to carry the maximum amount of traffic from all its customers at the same time.

In a circuit switched system, traffic that is not served because of lack of resources in the network is blocked and lost. The call originator must redial the number and hopefully succeed. In a store and forward system, e.g., in a packet switched system, if the server is not available, the incoming request is simply stored until it is available and then transmitted. In either case, the network is deemed to be congested. The impact of congestion reflects in the quality of service offered by the network.

The congestion phenomenon experienced by the user is mathematically captured, generally speaking, by the Erlang B formula in circuit switched systems as shown in the following section. Note that the Erlang B formula is nothing but the probability that all the n channels in an n-channel system are busy.

One measure of the quality of service in a circuit switched network, called the grade of service, is the ratio, unsuccessful call attempts divided by the total number of call attempts. The Erlang B formula is thus the grade of service.

6.2 Erlang B Traffic Characterization

The derivation of the Erlang B formula is based on two statistical assumptions characterizing the traffic. These are as follows: the arrival process of traffic is a Poisson process and the holding time (or the duration of a call in a telephone system) is exponentially distributed. These two assumptions are strong assumptions and may not always hold in practice. Even so, the observed results of actual implementations have been found to be in close correlation with the computations based on the use of the Erlang B formula. Additionally, since the Erlang B formula offers a closed form solution, it offers valuable insight into system behavior. For a derivation of the Erlang B formula, the reader can refer to any standard text on telecommunications traffic [1, 2]. The Erlang B formula is as follows:

$$E_n(A) = \frac{\frac{A^n}{n!}}{\sum_{k=0}^{n} \frac{A^k}{k!}} \tag{6.4}$$

$E_n(A)$ is the probability of blocking for the incident traffic A in Erlangs and n is the number of trunks or equivalently, the number of wavelengths in a dense wavelength division multiplexing (DWDM) system.

We now consider a network of arbitrary topology consisting of n nodes. We further define the associated parameters with the network in Table 6.1.

An observed characteristic of telecommunication networks is the reduction in throughput or delivered traffic when the network is congested. This is frustrating to

Table 6.1 Notations

n	Number of nodes in the network
w_{ij}	Number of available wavelengths on the link between two adjacent nodes i and j ($w_{ij} \geq 1$)
A	The total incident traffic intensity in Erlangs
$a_{sd}A$	The intensity of traffic incident on node s, destined to node d. Obviously, $\sum_s \sum_d a_{sd} = 1$
$l_{ij}A$	The (incident) traffic intensity on the link between adjacent nodes i and j.
p_{ij}	Blocking probability of the link between adjacent nodes i and j
p_{sd}	Blocking probability of the route from source node s to destination node d
C	The total traffic carried by the network, $A \geq C$
C_{sd-h}	Traffic carried between the src-dest pair s and d with h hops. Obviously $\sum_h C_{sd-h} \leq a_{sd}A$

the user as well as to the network service provider whose revenue depends not on the volume of service demands but on the volume of service delivered.

As a central office engineer, the senior author had the opportunity to observe this phenomenon several decades ago actually taking place. On busiest days when the number of call attempts peaked, the number of calls completed was often lower than on other days. A known "trick" to increase the call completion rate (and correspondingly, enhance customer satisfaction while also increasing the telephone company's take) was to manually kill calls that were still moving up in the switching hierarchy or, in other words, preferentially advantaging the calls that were closer to their destination and descending the hierarchy. A general take from the observation was that a reduction in the number of call requests from entering the network resulted in enhancing the number of successfully completed calls.

The phenomenon mentioned above is by no means unique to telecommunication networks. One way to reduce vehicular gridlocks in metropolitan areas is to limit the number of vehicles attempting to enter the gridlock prone part of the road system during a busy period. A traffic light that would actually limit the number of vehicles entering a freeway and activated during peak hours has been set up in several metropolises in order to limit the number of vehicles in the busy core and thereby enhance the movement of traffic.

A mathematical construct that would explain the phenomenon specifically for telecommunication networks was not available until recently. The construct presented below also offers insight into the design of networks, especially large, multi-hop networks. We first state and prove two theorems.

Theorem 6.1 *For a network of arbitrary topology, as the incident traffic intensity increases, the carried single-hop traffic between source-destination node pair increases until it reaches a finite limit [3].*

Proof The blocking probability of the link between adjacent nodes i and j is

$$p_{ij} = E_{w_{ij}}(l_{ij}A) = \frac{\frac{(l_{ij}A)^{w_{ij}}}{w_{ij}!}}{\sum_{k=0}^{w_{ij}} \frac{(l_{ij}A)^k}{k!}} \tag{6.5}$$

The carried single-hop traffic between adjacent node i and node j is

$$C_{ij-1} = a_{ij}A(1 - p_{ij}) = \frac{a_{ij}A + \cdots + \frac{a_{ij}l_{ij}^{(w_{ij}-1)}A^{w_{ij}}}{(w_{ij}-1)!}}{1 + \cdots + \frac{(l_{ij}A)^{w_{ij}}}{w_{ij}!}} \tag{6.6}$$

It can be easily seen from Eq. (6.6) that

$$\lim_{A\to\infty} C_{ij-1} = \frac{w_{ij}a_{ij}}{l_{ij}} \tag{6.7}$$

From Eq. (6.7), we can see that as incident traffic intensity increases, the carried single-hop traffic goes to a limit, which is given in Eq. (6.7). Note that $l_{ij}A$ may not generally be characterized as Poissonian even when $a_{st}A$ is Poissonian. Equation (6.5) is thus an approximation.

The outcome of Theorem 6.1 is also intuitive. Single-hop traffic in a network is comparable to traffic between the two end points of the network directly connected by a transmission facility. Since the facility can carry traffic to its capacity, but no more, it is obvious that it will reach a point of saturation beyond its capacity. A corollary of Theorem 6.1 is that if all the traffic in a network were single-hop traffic, there would be no multi-hop traffic consuming internal resources within the network. Such a network would be ideal from a resource utilization standpoint.

In practice, however, multi-hop networks are a reality because it is generally too expensive to have a n-node fully interconnected network with $n(n-1)/2$ links and each link having a capacity that would ideally fit the anticipated demand between the two end points. It is entirely possible, for example, that two nodes in the network remotely placed from each other have so little traffic as to justify a transmission facility directly connecting them.

The corresponding analysis for the case of a multi-hop network is considered next.

Theorem 6.2 *For a network of arbitrary topology, as the incident traffic intensity increases, the carried multi-hop traffic(i.e., traffic with two or more hops) between any source-destination node-pair increases, but eventually goes to zero, after reaching a peak [3].*

Proof Let nodes s, l, $m \ldots n$ and d be along the route from s to d. The blocking probability from node st to node d is

$$P_{sd} = 1 - (1 - p_{st})(1 - p_{tm}) \ldots (1 - p_{nd}) \qquad (6.8)$$

The carried multi-hop traffic from node s to node d is

$$
\begin{aligned}
C_{sd-h} &= a_{sd} A (1 - p_{sd-h}) \\
&= a_{sd} A (1 - p_{st}) \ldots (1 - p_{nd}) \\
&= a_{sd} A \frac{\sum_{k_1=0}^{w_{st}-1} \frac{(l_{st} A)^{k_1}}{k_1!}}{\sum_{k_1=0}^{w_{st}} \frac{(l_{il} A)^{k_1})}{k_1!}} \ldots \frac{\sum_{k_h=0}^{w_{nd}-1} \frac{(l_{nd} A)^{k_h}}{k_h!}}{\sum_{k_h=0}^{w_{nd}} \frac{(l_{nj} A)^{k_h})}{k_h!}} \\
&= \frac{a_{sd} A + \ldots + d_1 A^{w_{st}+w_{tm}+\ldots+w_{nd}-h+1}}{1 + \ldots + d_2 A^{w_{st}+w_{tm}+\ldots+w_{nd}}}
\end{aligned}
\qquad (6.9)
$$

For a given topology and traffic matrix, d_1 and d_2 are constants. Since $h > 1$, for multi-hop traffic, we have

$$\lim_{A \to \infty} C_{sd-h} = 0 \qquad (6.10)$$

Also, according to Rolle's Theorem [4], C_{sd-h} must reach a peak before it goes to zero as A tends to infinity. So, with increasing traffic intensity, the served multi-hop traffic also increases before it reaches a peak. If the traffic intensity continues to increase, the served traffic drops to zero. It is, therefore, obvious that with increasing traffic intensity, the multi-hop traffic continues to consume network resources without a commensurate increase in served traffic. As a corollary, we can observe that, for an arbitrary network topology, when the incident traffic intensity increases indefinitely, only single-hop traffic can be carried by the network. This can be seen as follows:

Proof

$$\lim_{A \to \infty} C = \lim_{A \to \infty} (C_1 + C_2 + \cdots + C_{h_{max}}) = \lim_{A \to \infty} C_1 \qquad (6.11)$$

since $\lim_{A \to \infty} C_h = 0$, when $h \geq 2$.

Theorem 6.2 and the accompanying corollary have shown that a multi-hop network can carry only single-hop traffic when traffic intensity becomes arbitrarily large. This phenomenon also offers an insight into the management system of a network under catastrophic conditions. Since most network systems can identify the destination at the point of origin of a demand for service, the system can actually kill originating traffic at any node unless it is destined for a node with direct link. In this case, the network can still keep functioning at its maximum capacity. The downside, of course, is that such a treatment might not be "fair" to all customers. On the other hand, it is possible to allow a small fraction of incident multi-hop traffic at the cost of network efficiency.

6.3 Summary

This chapter has presented an analytical basis for the intuitive inference that in order to maximize the traffic delivered by a network, the number of hops should be minimized. A large number of hops would result in traffic consuming bandwidth within the network without a corresponding increase in throughput. A well-designed network, in other words, should be as flat as possible.

Problems

6.1 Consider a system following the Erlang B model of traffic. The system has six trunks, i.e., $n = 6$. The incident traffic is 2 Erlangs. Compute the probability of blocking.

6.2 In the question above, assume the incident traffic increases by a factor of 3. To cope with the additional traffic, the number of trunks is increased by the same factor. Would the probability of blocking increase, decrease, or remain constant? Why?

References

1. D. Zwillinger, *Standard Mathematical Tables and Formulas*, 30th edn. (CRC Press, Boca Raton, 1996), pp. 335–335
2. V.B. Iverson, *Teletraffic Engineering and Network Planning* (Technical University of Denmark, DTU Fotonik, 2015)
3. Y. Qu, P.K. Verma, Limits on the traffic carrying capacity of optical networks with an arbitrary topology. IEEE Commun. Letters **8**(10), 641–643 (2004)
4. A. Birman, Computing approximate blocking probabilities for a class of all-optical networks, in *Proceedings of the IEEE, 14th Annual Joint Conference INFCOM'95*, vol.2 (1995), pp. 651–658

Chapter 7
Bandwidth and Throughput of Networks: Packet Switched Networks

7.1 Delay in Packet Switched Networks

The legacy circuit switched network allocates dedicated bandwidth for the duration of the call. Because dedicated bandwidth is available in circuit switched systems, the end-to-end delay is essentially fixed and not controllable. In contrast, packet switched systems undergo variable delays at each switching point during their journey because of the queue at the point. The delay we address in this chapter is this variable queuing delay.

Figure 7.1 presents a one-hop packet switched communication system. Packets arrive at point A at a rate of λ packets per second. The server transmits the packets toward their destination B at μC packets per second. During periods of time when the server is busy, i.e., in the process of transmitting a packet, an incoming packet will simply remain queued in the buffer shown in Fig. 7.1. For the system to be stable, μC must, of course, exceed λ. The ratio $\lambda/\mu C$ is termed the utilization of the system and is denoted by ρ. Therefore, $\rho < 1$. Note that C is the transmission speed of the line in bits/s and $1/\mu$ is the mean length of packets in bits.

Delay affects any interactive communication or any real-time application. Variations in delay, even within the bounds of acceptable maximum delays, in interactive applications such as Voice over IP applications, are especially unacceptable because of the negative impact of delay variations on the subjective quality of voice communication.

For variable delays, average delay is one measure of the quality of communication. This measure, of course, falls far short in defining the quality of service for human communication. For example, two systems A and B might have the same average delay, say 250 ms. However, if system A has delays ranging from 100 ms to 400 ms, while B has delays ranging from 200 to 300 ms, System B is much more suitable for human communication than the System A. A variation in delay causes jitter. In many ways, jitter is potentially the largest single contributor to degradation in the quality of human communication. One way to control jitter is through the

© Springer Nature Switzerland AG 2020
P. Verma, F. Zhang, *The Economics of Telecommunication Services*, Textbooks in Telecommunication Engineering, https://doi.org/10.1007/978-3-030-33865-7_7

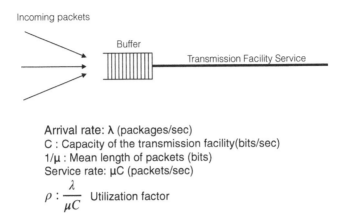

Incoming packets

Buffer

Transmission Facility Service

Arrival rate: λ (packages/sec)
C : Capacity of the transmission facility(bits/sec)
1/μ : Mean length of packets (bits)
Service rate: μC (packets/sec)

$\rho : \dfrac{\lambda}{\mu C}$ Utilization factor

Fig. 7.1 One-hop packet switched communication system

use of a jitter buffer which basically delays packets that arrive sooner than some anticipated delay. A reduction in jitter is thus achieved at the cost of higher average delay. The latter (higher average delay) will again reduce the quality of interactive communication.

For circuit switched systems where the delay was constant, it was easy to define throughput as we discussed in Chap. 6. But, for even a simple packet switched system shown in Fig. 7.1, it is difficult to define throughput because, if the buffer size is infinite, then, as long as the utilization factor is less than 1, all the incident packets will get delivered, although they might suffer an indefinitely large delay.

If the buffer is finite then, of course, some packets will drop off and be lost giving us a measure of throughput in terms of traffic that was actually delivered. However, modern technology has reduced the cost of buffer to such a low value that there is little motivation to reduce buffer sizes. And even in moderate size buffers, the loss of packets due to buffer overflow, generally speaking, is pretty low, making it an ineffective measure as a quality of service parameter for communication.

As in [1], we propose to define an upper bound on delay as an indicator of effective throughput for which a customer would be willing to pay. Stated another way, packets that suffer delays above a predefined threshold, willingly agreed to between the customer and the service provider, would be considered lost and would not constitute communication for which the customer would be liable to pay. Indeed it is entirely possible that service provider not only loose revenue for the packets delayed above the threshold, but also further compensate the user at some multiple of the lost volume of communication.

We do understand that this measure is not perfect for a variety of reasons. First, in any communication, not all packets or all segments of information are equally meaningful. For example, in human communication, a high degree of redundancy in any natural language permits understanding of the meaning even with small gaps and breaches in the communication system. Even so, putting up a bar on the delay, where an upper bound on the delay (rather than the average delay) is prescribed as a quality of service parameter, is a significant step in the right direction. It adds precision in the design of the communication system allowing the service provider to better estimate the resources needed and allows the user to make an informed choice on the suitability of a network for a specific application.

7.2 Analytical Model

Consider a LAN shown in Fig. 7.2 with a server. The LAN, as an example, functions as a packet switched network with callers and callees as the endpoints on the LAN. We further assume that the terminals on the LAN originate calls that are random; in other words, the distribution of the arrival time of the packets is Poissonian. We further assume that the distribution of the lengths of the packets is exponential. Packets that originate when the server is busy will wait until the server is free, i.e., when the transmission resource is available to transfer the packet. The server handles packets in the queue on a first in first out basis.

In this case, the distribution of the waiting time W can be written as [1]

$$F_w(t) = P\{W \le t\} = 1 - \frac{\lambda}{\mu C} e^{-(\mu C - \lambda)t} = 1 - \rho e^{-\mu C(1-\rho)t} \tag{7.1}$$

The throughput γ of the single-hop system in packets/s can be expressed as

$$\gamma = \lambda P\{W \le t\} = \lambda[1 - \frac{\lambda}{\mu C} e^{-(\mu C - \lambda)t}] \tag{7.2}$$

Equation (7.2) merits a close scrutiny. The term in the parenthesis

$$1 - \frac{\lambda}{\mu C} e^{-(\mu C - \lambda)t} = 1 - \rho e^{-\mu C(1-\rho)t} \tag{7.3}$$

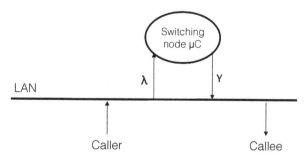

Fig. 7.2 Single-hop network

is positive, and less than 1, making γ or the throughput always less than λ. In other words, the number of packets delivered is always less than the number of packets entrusted to the network for delivery except when $t \to \infty$. For $t \to 0$, the number of packets delivered is arbitrarily small.

If we impose a finite value on t, i.e., count only those packets which arrive within the window of interval between 0 and t as constituting throughput, we have to accept the possibility of loosing a fraction of packets.

We now ask the following question: Assuming that the value of t is predefined for a certain application, is it possible to find an optimum value of λ which will maximize the throughput? Stated another way, if the traffic parameters λ, μ and the maximum allowable delay for constituting delivered traffic are known, what is the minimum requirement of the system capacity C? This latter question is extremely important for a service provider who must optimize the resources needed consistent with profitability.

The answer to the questions posed in the last paragraph is contained in the following theorem.

Theorem 7.1 *The throughput γ of a single-stage packet switched server, where the arriving traffic follows the M/M/1 discipline and all packets incurring a queuing delay higher than t are discarded, is maximized for a mean packet arrival rate λ_0 such that the following transcendental equation [1]:*

$$\lambda_0(2 + \lambda_0 t) = \mu C e^{(\mu C - \lambda_0)t} \tag{7.4}$$

is satisfied. The maximized throughput under this condition is given by

$$\gamma_{max} = \frac{\lambda_0(1 + \lambda_0 t)}{2 + \lambda_0 t} \tag{7.5}$$

Proof We note from Eq. (7.2) that γ is continuous on the closed interval $[0, \mu C]$. Thus, if γ_{max} is an extreme value of γ corresponding to λ_0 on that interval, then one of the following two statements is true: (a) $\gamma'(\lambda_0) = 0$, or (b) $\gamma'(\lambda_0)$ does not exist. The first-order derivative of Eq. (7.2) is

$$\frac{\partial \gamma}{\partial \lambda} = 1 - \frac{\lambda}{\mu C}(2 + \lambda t)e^{-(\mu C - \lambda)t} \tag{7.6}$$

which exists. The second-order derivative of (7.2) is

$$\frac{\partial^2 \gamma}{\partial^2 \lambda} = -\left[\frac{2(1 + \lambda t)}{\mu C} + \frac{\lambda t(2 + \lambda t)}{\mu C}\right]e^{-(\mu C - \lambda)t} \tag{7.7}$$

which is negative. We can now obtain the maximum throughput γ_{max} by putting the first derivative of γ, equal to zero. In other words, γ will be maximized for the specific λ_0 such that

$$1 - \frac{\lambda_0}{\mu C}(2 + \lambda_0 t)e^{-(\mu C - \lambda_0)t} = 0 \tag{7.8}$$

Equation (7.8) can be rewritten as

$$\lambda_0(2 + \lambda_0 t) = \mu C e^{(\mu C - \lambda_0)t} \tag{7.9}$$

The corresponding maximum throughput γ_{max} can be computed from Eq. (7.2) and Eq. (7.9) as

$$\gamma_{max} = \frac{\lambda_0(1 + \lambda_0 t)}{2 + \lambda_0 t} \tag{7.10}$$

Thus proves the Theorem.

Figure 7.3 shows plots of the throughput γ for varying levels of the incident traffic λ. The capacity of the server μC and t are used as parameters. It can be seen that given a fixed μC, the served traffic increases as the threshold of delay time t increases. Also, for a fixed t, the served traffic increases as the service rate μC increases. Both these results are also intuitive. A large service rate should obviously increase the throughput. Also, if there is a higher tolerance to delay, a higher level of throughput will have suffered delay lower than the specified threshold.

We note that for small values of the incident traffic λ, the delay suffered by the packets will be small and, from Eq. (7.2), the effective throughput γ will be nearly equal to λ. As λ increases, γ will increase even though a larger fraction of incident packets will not constitute throughput. Beyond a certain value of λ, however, the

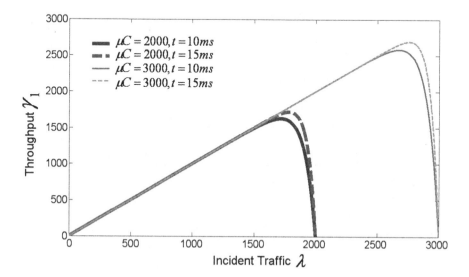

Fig. 7.3 Single-hop network throughput [1]

throughput will fall because a sharply increasing fraction of traffic will suffer delays larger than t and will, therefore, not constitute throughput.

7.3 Multi-hop Packet Switched Networks

We do see from Eq. (7.2) that the throughput γ for the case of a single-hop packet switched communication system can asymptotically approach the arrival rate as the threshold delay t increases indefinitely. Such is not the case with a two-hop or a multi-hop network. A two-hop network is shown in Fig. 7.4. The analysis of the two-hop network is encumbered by the fact that the traffic flowing out of node 1 toward node 2 is not necessarily Poisson distributed. However, under certainly idealized conditions [1, 2], the Poisson assumption has been shown to be valid and demonstrated to be adequate in practical situations. We omit the analysis here and refer the reader to [1] where it is shown that

$$P(W \leq t) = \rho^2 \left\{ 1 - e^{-\mu C(1-\rho)t} [1 + \mu C(1-\rho)t] \right\} \tag{7.11}$$

as the probability distribution function of the total delay.

As before, we have

$$\gamma = \lambda P(W \leq t) = \lambda^3 (\frac{1}{\mu C})^2 \left\{ 1 - e^{-(\mu C - \lambda)t} [1 + (\mu C - \lambda)t] \right\} \tag{7.12}$$

Further analysis leads to [1]

$$\gamma_{max} = \frac{\lambda_0^4}{3} (\frac{t}{\mu C})^2 (\mu C - \lambda_0) e^{-(\mu C - \lambda_0)t} \tag{7.13}$$

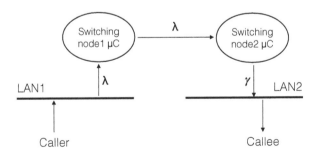

Fig. 7.4 Two-hop packet switched communication system

This is an important result pointing to the fact that two-hop networks are necessarily punitive in terms of delivering traffic under a specified end-to-end delay constraint.

If one were to necessarily push traffic through more than one hop for reasons of economic or administrative efficiency, the price to be paid will be through addition of capacity which would result in additional cost to the service provider.

The results can be extended to multi-hop networks which, indeed, has been carried out in the literature. We simply note here that as the number of hops increases the maximum throughput reduces, somewhat sharply, thus requiring the service provider to correspondingly increase capacity.

7.4 Impact of Multi-Hopping on Throughput

Chapters 6 and 7 have addressed the deleterious impact of multi-hopping on throughput for circuit switched and packet switched networks, respectively. In the case of circuit switched networks, the delivered traffic is sharply reduced as a function of network utilization beyond a certain point. In the case of packet switched networks, the total end-to-end delay rapidly increases thereby reducing the throughput because traffic that suffered delay more than a defined threshold is not considered to constitute throughput.

The primary reason for having traffic suffers a multi-hop journey between the source and the destination is that it is often cost-prohibitive to have a direct connection between all source-destination pairs. On the other hand, multi-hop traffic unquestionably involves a higher amount of resource consumption on part of the service provider for offering the same grade or quality of service. The material presented in Chaps. 6 and 7 offer the means to compute the additional amount of resource provisioning that must be made if the quality was to remain the same.

7.5 Summary

This chapter has offered an analytical solution to estimating the bandwidth necessary for packet switched networks under specified maximum delay constraints. The results obtained can be used to compute the maximum traffic bearing capacity of a packet switched network. Specifically, the chapter has brought out the negative impact of transiting through several hops between the source and the destination. The results obtained can lead to designing an optimal topology of a network and the bandwidth associated with each link depending on the customer preference for performance and pricing in a competitive market.

Problems

7.1 Throughput of a packet switched network in Chap. 7 has been defined differently from the throuhput for circuit switched networks discussed in Chap. 6. Explain the rationale for making this distinction.

7.2 If you were to adopt a different measure of throughput in a packet switched system what would it be?

References

1. P. Verma, L. Wang, *Voice over IP Networks* (Springer, New York, 2011)
2. Kleinrock, L., Queueing systems, in *Theory*, vol. 1 (Wiley, New York, 1975)

Chapter 8
Pricing of Telecommunication Services

8.1 The Role of Pricing in Network Services

Pricing plays an important role in any business endeavor. In competitive businesses, pricing decisions are based on several tactical and strategic considerations. Telecommunication network services pricing is based on several additional considerations. One example: Network services cannot be stored and dispensed later, unlike most commodities. Furthermore, network services are subject to regulations by the central and local government related to pricing and geographic availability in most countries, including the USA.

We have seen in Chaps. 6 and 7 that congestion takes place as the demand for services goes beyond the capacity of the network. Past this point, any increase in the incident traffic results in progressively reducing throughput. This phenomenon increases customer frustration while lowering revenue of the service provider at the same time. Therefore, shaping users' demand plays an important role in pricing telecommunication services. The operational cost of a telecommunication network is largely invariant as a function of demand. This makes it desirable to have the network operate at its rated capacity most of the time, if at all possible. Pricing is one mechanism that can accomplish this.

Pricing has one other major impact. Low prices stimulate demand and, conversely, higher prices suppress demand. However, in a competitive environment, high prices also stimulate capacity supply by emerging competitors based on the prospect of higher profitability. When the market is in equilibrium, the supply of network services closely matches the demand. Low prices, on the other hand, are likely to result in unmet demand as fewer suppliers would likely enter the market. Excessive pricing, on the other hand, will likely have unused network services capacity.

In the absence of congestion, the marginal cost of telecommunication network service is close to zero. Stated in another way, the network service provider does not incur any additional cost to convey information if the network is not congested.

© Springer Nature Switzerland AG 2020
P. Verma, F. Zhang, *The Economics of Telecommunication Services*, Textbooks in Telecommunication Engineering, https://doi.org/10.1007/978-3-030-33865-7_8

In the case of commodities, competition tends to drive prices toward the marginal cost of the commodity. Telecommunication network services are a far cry from this for two reasons. First, even though the incremental cost of serving traffic is zero, continual increase in traffic will result in severe degradation or potential collapse of the network at some point. Second, while traffic increases incrementally, network capacity does not, as it might require major costly upgrades.

Regulatory framework prevalent in most countries further constrains the pricing flexibility of a network services provider. In addition to market factors driven by consumers and suppliers, regulatory factors thus become an important consideration in setting prices and, consequently, guiding investment decisions in telecommunication networks. Chapter 11 addresses the impact of regulation on pricing.

Telecommunication network service customers want predictable and transparent pricing schemes along with the ability to audit their usage on demand. Generally speaking, telecommunication network service users have also expressed their inclination to have a flat rate pricing scheme. This obviates the risk of unpredictable total charges at the end of the month. In many other situations, such as in metered electricity usage service, users feel empowered to control the usage. The same is not the case in telecommunication services usage since telecommunication is a bilateral exchange. The service originating user (and the one who would pay for the connection charges) is not always in control of the duration of the communication or the volume of information transferred. Overall, pricing of telecommunication services requires a multitude of considerations many of which transcend the pricing considerations of other commodities and services.

8.2 Traditional Methods of Pricing Telecommunication Services

Until about the mid-1980s, telecommunication services in most countries were monopolies and in most cases owned by the central government of the country. In the developed world, monopoly pricing has largely given way to market-based pricing schemes. The change has not, however, occurred in its entirety and regulatory governance is an important component of telecommunication services pricing mechanism in most countries, including the USA.

In a highly regulated, monopoly environment, telecommunication providers could maximize their profits by raising prices to inefficiently high levels at the expense of users' welfare and reduced demand for telecommunication services. Because of the traditionally high barriers to enter the telecommunication industry, the monopoly situation in telecommunication services tends not to be a temporary phenomenon. To protect users from monopoly pricing, federal and state policymakers enforce rate regulation on service providers.

Traditional methods of regulated pricing focused on controlling the prices to the customers while limiting the profitability of the service provider. Known as the rate of return regulation, this scheme allowed depreciation of the invested capital, but it also insured that the total profit of the service provider based on revenue and cost (including the cost of depreciated capital and the cost of operational expenses) did not exceed a mandated rate of return. The components that filled the basket of services and their individual pricing were left largely to the discretion of the service provider and the regulator simply required that the services as a whole did not make a profit beyond the allowed rate of return.

One would see that the service provider in such a scheme would have a propensity to indulge in investments with a long cycle of depreciation of capital investments while having little incentive to engage in activities that would reduce costs, either capital or operational. On the other hand, the scheme allowed the monopoly service provider to engage in long-term planning, and research and development activities, and not have their business decisions driven by transient behaviors of competitors or the customers. It offered stability of the business processes and predictability of the volume of business avoiding, by and large, disruptive practices.

The rate of return regulation (discussed more fully in Chap. 11) would give way, incrementally, to other regulatory practices and schemes. One such scheme was called, price cap regulation, which fixed the price, say, in year 1, based on rate of return regulatory practice. In the subsequent years, the price was determined by the cost trends in technology and general inflation, such as the inflation in wages. This scheme did not limit profitability and it encouraged the service provider to use newer technology which is generally cheaper and more functional.

One would anticipate that the actual practice of rate regulation based on either of the two regulatory schemes mentioned above would be fraught with varying interpretations of the market data by the personnel engaged by the service provider and the regulator. This was actually the case as observed through massive and often fractious regulatory hearings and the eventual coming to terms at some level between the extreme positions of regulator and the service provider.

The first opening and movement toward competitive pricing occurred in the case of Internet pricing. The Internet was seen as an information service as opposed to telecommunication service, offered by a plurality of service providers in a competitive environment. It was exempted from price regulation. With the growth in the Internet services, which would eventually encompass all aspects of telecommunication services, the legacy regulatory practices continue to be deemphasized leading possibly to their eventual extinction.

8.3 Characteristics of Communication Services

When a communication network is built, the infrastructure cost is largely a fixed cost. The variable operating cost is extremely small compared to the fixed cost. As is well known, the first copy of a software product incurs the cost of development. The subsequent copies are cheap to produce. All additional copies can be produced at almost zero marginal cost. Similarly, once a network is built, the marginal cost of providing a unit of communication service can be almost zero, especially when there is no congestion. And it is well known that a competitive market drives prices toward marginal costs. As a result, there is a danger for the communication industry that the prices of communication services can be driven close to zero.

Returning to the subject of communication services, it should be noted that they can sell at both low and high prices. For example, there are hundreds of web sites providing email services, and it seems they cannot charge users because there are many nearly equivalent sites providing similar service. Such a service is termed a *commodity service*, which has little pricing flexibility. Providers of a commodity service would find it more profitable to concentrate on differentiating their services by providing value-added features like security for which some users may be willing to pay. Besides the fact that different applications require different QoS, service differentiation will allow a service provider to not be perceived as a commodity provider. The reason is that if goods are not a commodity, it can sell at a price that reflects its value to users, rather than its production cost, that is, its marginal cost.

From a regulatory perspective, it can be noted that as long as network operators offer equal treatment to users' traffic within each class of transport service, the service differentiation across classes might not violate the net neutrality [2].

A special feature of communication services is their reliance on statistical multiplexing. This is because data traffic is often bursty and sporadic, and not all users are active at the same time. Therefore, statistical multiplexing produces economy of scale. Thus the size of the user base can increase more than proportionately to the level of the network resource needed for identical performance. Besides network externality discussed in Chap. 1, statistical multiplexing is also an incentive for the network provider to price its services attractively so as to expand its user base.

Another special feature of the communication services is that the performance obtained by any network user is not only determined by his own traffic and service choice but also by other users' traffic and service choices. This is true for Internet best-effort (single grade QoS) networks; however, in multi-service networks this interdependence is much more complex because priority service traffic may cause performance variation to all others, even when the aggregate traffic load remains constant. This interdependence among users can be addressed through a game-theoretic framework in which each user makes a service choice and traffic volume or throughput rate that maximizes his or her utility while taking into consideration all the other users' choices. Such a delicate balance is achievable by applying the principles of game theory.

In this section, we have discussed features of telecommunication services: the marginal cost of the communication services is close to zero, and in order to recover the huge sunk cost and be profitable, service differentiation provides a potential avenue. Statistical multiplexing produces economy of scale for communication networks; the interdependence among users from a usage and pricing perspective can be effectively addressed using a game-theoretic framework. Game theoretic approaches are discussed in Chap. 12 and subsequent chapters.

8.4 Other Considerations in Pricing Network Services

One obvious role of pricing is for network providers to recover the capital investment and generate extra revenue to become profitable. As discussed in Chap. 1, the network externalities effect prompts network providers to increase the value of the network by reducing price to attract more demand. Network providers can also increase prices to control congestion and smoothen bursty customer demand. Therefore, pricing can be viewed as a control mechanism to shape users' demand.

An alternate view is that future networks will be overprovisioned [3]. By taking advantage of the reduced cost of new technologies, overprovisioning can solve the problem of congestion, and network providers do not need to use pricing as a mechanism to control the volume of network traffic. Overprovisioning might be reasonable for the backbone of the network because it consists of a fairly small number of links. But there is substantially less fiber installed in the access part of a network, which connects users to the core network. The core network infrastructure is shared by all users, but the access network is used by much fewer users. It is generally believed that it would take 20–30 times as much time and expense to overprovision an access part of the network as it has taken to build the fiber infrastructure in the backbone [1]. Although AT&T and Verizon have the fiber to home services like Uverse and FIOS, respectively [4], these services are only provided in a limited number of metropolitan cities.

In addition, it is always hard for any network operator to predict demand. Just as it was overestimated in the 1990s, it may now be underestimated. There are increasing amounts of traffic on the network generated by programs and devices connected to the Internet, rather than by humans. These programs and devices can ultimately greatly outnumber human users. Thus, network traffic has the potential to grow extremely rapidly. The demand for telecommunication network services is constantly in a transient phase. On the other hand, the expansion of capacity of the network can only be done in sizable steps. Pricing plays an important part in matching the demand to supply. Without effective pricing schemes, a network provider will not have enough incentive to invest, and, therefore, overprovisioning may never happen in the marketplace. Without enough network capacity, Internet innovations will be restricted and will adversely affect users' experience. Thus, in addition to being a control mechanism, pricing is also an important factor in keeping the information industry economically viable.

Pricing can produce the right incentives for users to choose levels of service in service differentiated networks that are most appropriate for them in order to ensure that resources are not wasted. The latency requirements of email services and voice over IP applications are different. Accordingly, their QoS requirements are different. Appropriate level of pricing for these services will be important for effective use of resources. While the proposed implementations vary in different studies [5–10], the basic idea is that the appropriate pricing policy is an important tool for controlling user demands while ensuring effective utilization of network resources at an appropriate level.

Pricing can be further viewed as a signal from the network operator to its user base that there are incentives to use the network efficiently. This is very similar to the Transmission Control Protocol (TCP) in the Internet. When TCP receives a congestion signal from the network, it reduces the sending rate; otherwise it increases the sending rate. However, a user might be motivated to override the TCP in order to increase the rate of transmission. This can be avoided if the congestion causing traffic were to be charged more [11]. Use of this contrivance provides stability and robustness to the network.

8.5 Related Work on Pricing Network Services

8.5.1 Pricing for Regulated Telecommunication Services

Providers of services like telephony offered in the Public Switched Telephone Network (PSTN) are considered common carriers. Pricing of such services is generally subject to regulation. Traditionally, there has been very little competition in such markets. With the advent of Internet services like VoIP, email, etc., the marketplace for these services has changed much; however, it is still interesting to review the traditional regulatory pricing scheme for network operators.

Like any other public utility, an incumbent telecommunication service provider has made a regulatory pact with the government in which the company is given an opportunity to earn a "reasonable rate of return" on its overall regulated investment. Under a rate of return regime, federal and state regulation gives dominant incumbents opportunities to charge retail rates sufficient to cover their anticipated expense plus a reasonable return on their net investment [12].

8.5.2 Pricing Internet Services

Internet access is an information service and not a telecommunication service. It is thus exempt from telecom regulation [13]. In this non-monopolized market, network providers compete against one another for users, and this competition theoretically keeps the price of service at reasonably efficient levels. The most widely used pricing schemes for the Internet services include access-rate dependent charges, usage-dependent charges, or a combination of both [14]. An access-rate dependent charge has the following two forms: unlimited use, or limited time of the connection and charging a per-minute fee for additional connection time. Similarly, the access and usage-dependent charging scheme allows a fixed access fee for a defined usage to be transmitted, and then imposes per-unit volume charge for additional use. A brief summary of the main advantages and disadvantages of flat rate and usage-dependent rate is presented in Table 8.1.

Table 8.1 Arguments for and against flat rates and usage-dependent pricing

	Advantages	Disadvantages
Flat rate	Easy to implement	Unfair to light users
	Little billing overhead	May lead to service overuse
Usage-dependent	Increased fairness	Adverse response from users
	Can be used for congestion control	Billing complexity
		Reduced usage

An access-rate dependent flat rate is the method used in the USA to charge for Internet use. Light users (e.g., email, occasional web browsing) may, therefore, subsidize the heavy users (e.g., multimedia applications, frequently downloading of large files) [15]. In addition, the unbridled consumption may lead to overuse of the network resources. However, because users have strong preference for flat rate pricing, despite the above disadvantages, service providers still stick to a flat rate model in order to avoid losing customers to a competitor [16].

Usage-dependent pricing can be a solution for the problem of fairness and service overuse. However, this policy makes it difficult for users to budget for a network expense, not only because it is hard for users to predict their own traffic statistics, but also because the Internet is an interactive experience and users are not fully in control of their usage. These evidences make Internet users not react favorably to a usage-dependent pricing scheme [17]. Furthermore, for network providers, the additional costs in billing may be substantial and must be offset by the gains brought by usage-based pricing. In traditional telephony, more than half of what users pay for a call goes to cover the cost of providers' accounting system [18], and this is in a circuit switched system in which there is no need to count how many packets traverse the network. Finally, usage-dependent pricing tends to discourage the use of the Internet which is in contrast to the network externality discussed in Chap. 1.

Well-known proposals for Internet pricing rely on a centralized optimization process to maximize the total users utilities [11, 19–22]. Kelly [11] forms a distributed flow control algorithm using the gradient ascent method from optimization theory which continuously informs the selfish users prices according to the network condition. Selfish users, who seek to maximize their own net benefit, are given the prices that have the right incentives to globally optimize the social benefits. An Explicit Congestion Notification (ECN)-based marking has been proposed in [23] to convey congestion information back to the end points. The resulting system converges to an optimal system state as long as all utility functions are strictly concave. Instead of only marking the packets during periods of congestion, [24] has proposed assigning each packet a price to reflect the congestion of the network. However, it is not clear whether all these theoretical results hold in the presence of transmission delay at the scale of a large network. In addition, all these schemes assume network services are best-effort and rely on a pure market mechanism to maximize social benefits.

The Internet, a single-service or best-effort service network, cannot support the performance needs of heterogeneous applications unless it is extremely overprovisioned [25]. Moving from a single-service to multiple-service architecture adds new dimensions to the pricing issue. It is obvious that a flat rate would no longer provide adequate incentives for users' choice of services, and therefore service-class-dependent, congestion-sensitive approaches must also be investigated. The next sub-section reviews some of the progress made in pricing of multi-service communication networks.

8.5.3   Recent Literature on Pricing Multi-Service Communication Networks

A number of articles have been published on telecommunication engineering and economics investigating the subject of pricing for multi-service networks. We summarize some of these studies.

Pricing based on network resource consumption has been considered in [26–31]. A study [26] has proposed a pricing algorithm in a DiffServ environment based on the cost of providing different levels of services and on long-term average user resource demand of a service class. The network service is dynamically priced based on the level of service, usage, and congestion-sensitive parameters. The study [30] has presented a mechanism that introduced a priority system with the objective of providing a higher and a lower quality of service to two customer groups. The non-priority traffic carries a lower price tag and a lower quality of service. An important characteristic of the proposed pricing schemes is that the overall revenue associated with the network would remain constant as long as the total demand is confined within a relatively large bound, termed the region of operation, for the network. Like the region of operation defined in [30], in order to make sure the prices for higher QoS are larger than prices for lower QoS, [26] also assumes the long-term demand for higher QoS traffic is lower than demand for lower QoS traffic. Reference [28] has proposed that users be charged a price per unit of effective bandwidth used. Assuming that the network knows its capacities and virtual path routing, as well as users' benefit function and traffic stream characterization, the paper has discussed the role of pricing in meeting users' needs, network resource allocation, and contract negotiation to form a complete connection provisioning process. Reference [29] has studied a network that offers its bandwidth and buffers for rent. The network periodically adjusts prices based on monitored user requests for resources with the objective of maximizing social welfare. Users reserve resources based on individual traffic parameters and delay requirements so as to maximize their utilities subject to budget constraints.

Microeconomic supply-demand principles have been applied to network traffic management problems. The studies in [32–34] rely on a centralized optimization process to maximize the total user utility. Kelly [32] has described a system in which users reveal how much they are prepared to pay per unit time. Then the network determines allocated rates so that the rate per unit charge are proportionally fair. The author has determined that the optimum system in this case is achieved when users' choices of charges and the network's choice of rates are in equilibrium. Reference [33] has studied the efficiency of using one bit to carry streams with differential QoS requirements in an attempt to maximize network revenue. In [35, 36], the resource is priced to reflect demand and supply. The method in [35] relies on well-defined source model and cannot adapt well to changing traffic demands, while the scheme in [36] also takes into account network dynamics (sessions join or leave) and source traffic characteristics and allows different equilibrium prices over different time periods. An economic equilibrium model is proposed in [34] which describes utility maximization by users and revenue optimization by service providers. In the presence of competing providers, the equilibrium prices reduce to the marginal costs. Study [37] has borrowed the framework described in [24] and calculates a price for each packet based on its bandwidth consumption, service level, and buffer occupancy. Reference [38] adjusts bandwidth and buffer allocations among classes to guarantee the target delay and loss.

Several studies have demonstrated through experiments or simulations that service-class sensitive pricing results in higher network performance. Reference [39] has proposed a Paris Metro Pricing (PMP) scheme which partitions the network into logically separated classes with different prices for each. It is expected that the higher-priced class will have less load and will provide better service. The behavior of PMP under equilibrium conditions is considered and compared with a uni-class pricing system in [40, 41]. Study [42] has analyzed the equilibrium of such a system using non-cooperative game theory. Reference [43] has considered a similar framework based on queuing theory and experiments. All of the above works consider the impact of differential pricing on the relative performance of the system as a result of user self-selection process. References [44, 45] have used simulations to study the problem of customer decisions in a two-priority network, where a fixed per unit price is associated with each priority class. These studies have concluded that, through the use of class-dependent pricing, it is possible to set prices so that all users are more satisfied with the resultant cost/benefit provided by the network.

Several opportunity cost-based mechanisms have been studied. Reference [46] has addressed the impact of QoS on bandwidth requirements in IntServ networks and proposes a scheme in which a service provider can develop compensatory and fair prices for users with varying QoS. Since exclusive allocation of bandwidth to a specific flow has a performance penalty on delay and jitter to other flows, [46] has derived the additional capacity required to maintain the desired performance of other flows, and has proposed a compensatory scheme that will fairly charge the specific flow requesting exclusive bandwidth. Reference [27] has developed a grade of service (GoS) based pricing scheme that results in efficient utilization of the network bandwidth and buffers. Essentially, each traffic is charged an amount of money based on the QoS degradation caused to other users sharing network resources. Price is, therefore, a function of the network utilization as well as individual utilities. Reference [47] has presented an approach based on the notion of cost in the context of providing services with differentiated levels of quality. In [47], the authors have investigated the impact of multiple traffic classes on the carrying capacity of a network with a prescribed threshold of blocking probability in a DWDM ring network architecture.

Auction-based mechanisms have been studied in [7, 48, 49]. The smart market model has been studied in [7], in which prior to transmission, users inform the network of how much they are willing to pay for the transmission of a packet; packets are admitted if their bids exceed the current cutoff amount, determined by the marginal congestion cost imposed by an additional packet. Users do not pay the price they actually bid, but rather the market-clearing price, which is always lower than the bids of all admitted packets. However, this mechanism only provides a priority relative to others and it does not promise quality of service. The Generalized Vickrey Auction (GVA) model in [49] supports multiple levels of QoS guarantees. But the optimal solution requires substantial computation, which increases as polynomial time with the number of users. The Progressive Second Price auction (PSP) scheme in [48] has extended the traditional, single, non-divisible object auction to the allocations of arbitrary shares of the total available resource with associated bids.

A set of game-theoretic analyses have been proposed for QoS provisioning and network pricing. In [50], packets are marked according to users' QoS requirements and the costs incurred to users are dependent on performance. The authors in [51] have investigated the dimensioning of network capacity for different service classes. References [52, 53, 55] have studied a static pricing scheme based on the priority classes. Reference [55] has described a method to predict each user's service choice in a game-theoretic framework given any price difference between services and an estimate of users' utility functions. Therefore, the service provider can determine the price ranges that encourage users to exhibit behavior that is beneficial to both users and providers. The work in [54] has generalized the idea in [7] to support auctions for different service levels.

8.6 Summary

This chapter has presented general considerations for pricing telecommunication network services. It has considered the generalized role of pricing as well as the legacy and emerging pricing practices for telecommunication services.

Problems

8.1 A general characteristic of all common user networks is that it has a problem in handling the peak demand. As a network service provider, how would you handle the problem?

8.2 Comment on the use of overprovisioning as a means to avoid congestion in telecommunication networks.

References

1. C. Courcoubetis, R. Weber, *Pricing Communication Networks: Economics, Technology, and Modelling, West Sussex, England* (Wiley, Hoboken, 2003)
2. T. Wu, Network neutrality, broadband discrimination. J. Telecommun. High Technol. Law **2**, 141 (2003). https://doi.org/10.2139/ssrn.388863. SSRN 388863
3. D. Isenberg, *Research on Costs of Net Neutrality* (2007). http://isen.com/blog/2007/07/research-on-costs-of-net-neutrality.html. Accessed Mar 2020
4. *Who Wins: Verizon FiOS vs AT&T U-Verse* (2008). http://gigaom.com/2008/08/19/who-wins-verizon-fios-vs-att-u-verse/. Accessed Mar 2020
5. A. Gupta, B. Jukic, M. Parameswaran, D.O. Stahl, A.B. Whinston, Streamlining the digital economy: how to avert a tragedy of the commons. IEEE Int. Comput. **I**(6), 38–47 (1997)
6. M.L. Honig, K. Steiglitz, Usage-based pricing of packet data generated by a heterogeneous user population, in *Proceedings of the IEEE INFOCOM*, vol. 2 (IEEE, Boston, 1995), pp. 867–874
7. J.K. MacKie-Mason, H.R. Varian, *Pricing the Internet, Public Access to the Internet* (JFK School of Government, Cambridge, 1993)
8. J.K. MacKie-Mason, H.R. Varian, Pricing congestible network resources. IEEE J. Sel. Areas Commun. **13**(7), 1141–1148 (1995)
9. C. Parris, S. Keshav, D. Ferrari, *A Framework for the Study of Pricing in Integrated Networks*. Technical Report. International Computer Science Institute, Berkeley, 1992
10. L. Murphy, J. Murphy, J.K. MacKie-Mason, Feedback and efficiency in ATM networks, in *Proceedings of the International Conference on Communications (ICC'96), Dallas* (1996), pp. 1045–1049
11. F. Kelly, A. Maulloo, D. Tan, Rate control in communication networks: shadow prices, proportional fairness and stability. J. Oper. Res. Soc. **49**, 237–252 (1998)
12. J. Nuechterlein, P. Weiser, *Digital Crossroads: American Telecommunications Policy in the Internet Age* (The MIT Press, Cambridge, 2004)
13. Federal Communications Commission, *New Principles Preserve and Promote the Open and Interconnected Nature of Public Internet* (2005)
14. P. Reichl, S. Leinen, B. Stiller, A practical review of pricing and cost recovery for internet services, in *Proceedings of the 2nd Internet Economics Workshop Berlin (IEW '99), Berlin)* (1999)
15. R. Edell, P. Varaiya, Providing internet access: what we learn from INDEX. IEEE Netw. **13**(5), 18–25 (1999)
16. J. Bailey, L. Mcknight, *Internet Economics: What Happens When Constituencies Collide, INET'95, Honolulu* (1995), pp. 659–666
17. L Mcknight, J. Bailey, Internet economics: when constituencies collide in cyberspace. IEEE Int. Comput. **I**(6), 30–37 (1997)
18. H. Brody, *Internet@crossroads*, Technology Review (1995)
19. C. Courcoubetis, V. Siris, Managing and pricing service level agreements for differentiated services, in *Proceedings of the IEEE/IFIP IWQoS'99, London* (1999)
20. P. Reichl, S. Leinen, B. Stiller, A practical review of pricing and cost recovery for internet services, in *Proceedings of the Second Internet Economics Workshop Berlin (IEW'99), Berlin* (1999)
21. J. Altmann, K. Chu, A proposal for a flexible service plan that is attractive to users and internet service providers, in *Proceedings of the IEEE INFOCOM, Anchorage* (2001)
22. S. Floyd, V. Jacobson, Link-sharing and resource management models for packet networks. IEEE/ACM Trans. Netw. **3**(4), 365–386 (1995)
23. R.J. Gibbens, F.P. Kelly, Resource pricing and the evolution of congestion control. Automatica **35**, 1969–1985 (1999)
24. A. Ganesh, K. Laevens, R. Steinberg, Congestion pricing and user adaptation, in *Proceedings of the IEEE INFOCOM, Anchorage* (2001), pp. 959–965

25. S.J. Shenker, *Service Models and Pricing Policies for an Integrated Services Internet* (1993). http://citeseerx.ist.psu.edu/viewdoc/summary?doi=10.1.1.38.6269
26. X. Wang, H. Schulzrinne, Pricing network resources for adaptive applications. IEEE/ACM Trans. Netw. **14**(3), 506–519 (2006)
27. H. Ji, J.Y. Hui, E. Karasan, GoS-based pricing and resource allocation for multimedia broadband networks, in *Proceedings of the IEEE INFOCOM, San Francisco* (1996), pp. 1020–1027
28. H. Jiang, S. Jordan, A pricing model for high speed networks with guaranteed quality of service, in *Proceedings of the IEEE INFOCOM, San Francisco* (1996), pp. 888–895
29. S.H. Low, P.P. Varaiya, A new approach to service provisioning in ATM networks. IEEE/ACM Trans. Netw. **1**(5), 547–553 (1993)
30. F. Zhang, P. Verma, A constant revenue model for packet switched network, in *IEEE GIIS 09, Hammamet* (2009)
31. H.R. Sukasdadi, P.K. Verma, A constant revenue model for telecommunication networks, in *International Conference on Systems and International Conference on Mobile Communications and Learning Technologies* (2006)
32. F.P. Kelly, Charging and rate control for elastic traffic. Eur. Trans. Commun. **8**, 33–37 (1997)
33. S. Ramesh, C. Rosenberg, A. Kumar, Revenue maximization in ATM networks using the CLP capability and buffer priority management. IEEE/ACM Trans. Netw. **4**(6), 941–950 (1996)
34. K. Kumaran, M. Mandjes, D. Mitra, I. Saniee, Resource usage and charging in a multi-service multi-QoS packet network, in *Proceedings of the MIT Workshop on Internet Service Quality Economics* (1999)
35. N. Anerousis, A.A. Lazar, A framework for pricing virtual circuit and virtual path services in ATM networks, in *Proceedings of the ITC-15* (1997)
36. E.W. Fulp, D.S. Reeves, Distributed network flow control based on dynamic competitive markets, in *Proceedings of the ICNP98* (1998)
37. A.J. O'Donnell, H. Sethu, A novel, practical pricing strategy for congestion control and differentiated services, in *Proceedings of the ICC* (2002), pp. 986–990
38. S. Jordan, Pricing of buffer and bandwidth in a reservation-based QoS architecture, in *Proceedings of the ICC* (2003), pp. 1521–1525
39. A. Odlyzko, Paris metro pricing: the minimalist differentiated services solution, in *Proceedings of the NOSSDAV'99, Basking Ridge* (1999)
40. R. Gibbens, R. Mason, R. Steinberg, Internet service classes under competition. IEEE J. Sel. Areas Commun. **18**(12), 2490–2498 (2000)
41. R. Jain, T. Mullen, R. Hausman, Analysis of Paris metro pricing strategy for QoS with a single service provider, in *Proceedings of the IWQoS* (2001), pp. 44–48
42. P. Marbach, Pricing differentiated services networks: bursty traffic, in *Proceedings of the IEEE INFOCOM, Anchorage* (2001), pp. 650–658
43. J. Altmann, H. Daanen, H. Oliver, A.S.-B. Suarez, How to market-manage a QoS network, in *Proceedings of the IEEE INFOCOM*, vol. 1 (2002), pp. 284–293
44. R. Cocchi, D. Estrin, S. Shenker, L. Zhang, A study of priority pricing in multiple service class networks. ACM SIGCOMM Comput. Commun. Rev. **21**(4), 123–130 (1991)
45. R. Cocchi, S. Shenker, D. Estrin, L. Zhang, Pricing in computer networks: motivation, formulation, and example. IEEE/ACM Trans. Netw. **1**(6), 614–627 (1993)
46. M.H. Dahshan, P.K. Verma, Pricing for quality of service in high speed packet switched networks, in *High Performance Switching and Routing Workshop, Poznan* (2006)
47. Y. Qu, P.K. Verma, Notion of cost and quality in telecommunication networks: an abstract approach. IEE Proc. Commun. **152**(2), 167–171 (2005)
48. N. Semret, A. Lazar, The progressive second price auction mechanism for network resource sharing, in *Proceedings of the 8th International Symposium Dynamic Games, The Netherlands* (1998)
49. J.F. MacKie-Mason, A *Smart Market for Resource Reservation in a Multiple Quality of Service Information Network*, Technical Report. University of Michigan, Michigan, 1997

50. S. Chen, K.I. Park, An architecture for noncooperative QoS provision in many switch systems, in *Proceedings of the IEEE/INFOCOM* (1999), pp. 865–872
51. P. Fuzesi, A. Vidacs, Game theoretic analysis of network dimensioning strategies in differentiated services networks, in *Proceedings of the ICC* (2002), pp. 1069–1073
52. M. Mandjes, Pricing strategies under heterogeneous service requirements, in *Proceedings of the IEEE/INFOCOM*, pp. 1210–1220 (2003)
53. P. Marbach, Analysis of a static pricing scheme for priority services. IEEE/ACM Trans. Netw. **12**(2), 312–325 (2004)
54. J. Shu, P. Varaiya, Pricing network services, in *Proceedings of the IEEE/INFOCOM* (2003), pp. 1221–1230
55. L.A. Dasilva, D.W. Petr, N. Akar, Equilibrium pricing in multi-service priority-based networks, in *Proceedings of the IEEE/Global Telecommunications Conference (GLOBECOM)*, vol. 3 (1997), pp. 1373–1377

Chapter 9
Pricing of Circuit Switched Services

9.1 Resource Consumption Based Pricing for Circuit Switched Networks

Chapter 6 has addressed the reduction in throughput for circuit switched multi-hop networks. If we were to keep the quality of service (measured as the probability of blocking) in a multi-hop network, bounded to a certain value, we must have more bandwidth in each component of the end-to-end transmission facility when compared with a single-hop network. In this section, we consider the additional amount of resources needed for multi-hop networks if the quality of service were to remain identical to that of a single-hop network with identical traffic intensity.

We consider a circuit switched network with "blocked and lost" or Erlang B characteristic. The network could be an optical network using DWDM (dense wavelength division multiplexing) or a legacy network. In the former, the transmission resource is a wavelength and, in the latter, it is a trunk. The quality of service is characterized by the probability of a call (or request for service) being blocked because of lack of transmission resource which could be either an unavailable wavelength or a trunk. As seen in Chap. 6, multi-hop traffic suffers a higher probability of being blocked compared to single-hop traffic independent of the topology of the network. Furthermore, when the network is congested, the throughput of the multi-hop traffic reduces to zero. Our analysis proceeds as follows.

With p as the blocking probability of a link, the probability that the link is available is $(1-p)$. In case there are h hops in tandem then, assuming that each of the transmission links has the same blocking probability, and the blocking probabilities of the individual transmission links are independent, the probability of availability of the end-to-end transmission system is $(1-p)^h$. The end-to-end blocking probability is, thus,

$$p_h = 1 - (1-p)^h \tag{9.1}$$

© Springer Nature Switzerland AG 2020
P. Verma, F. Zhang, *The Economics of Telecommunication Services*, Textbooks in Telecommunication Engineering, https://doi.org/10.1007/978-3-030-33865-7_9

Let us now consider that a network carries single-hop and multi-hop classes of service. In this case, the multi-hop traffic traversing the maximum number of hops in the network should suffer a probability of blocking no more than p.

From Eq. (9.1), it can be easily seen that the blocking probability for a h-hop system increases as h increases. From Eq. (9.1), it follows that

$$p = 1 - (1 - p_h)^{1/h} \tag{9.2}$$

Equation (9.2) can be interpreted as follows. If the end-to-end probability of blocking of a h-hop system were to remain the same as that of a single-hop system, the probability of blocking of each link in the h-hop system must correspondingly decrease.

We now compare the throughput of a single-hop system with that of a two-hop system assuming that the link capacities are identical. We keep the incident traffic A identical for the single and the two-hop systems.

For the single-hop system, the carried traffic C_1 is given by

$$C_1 = A(1 - p) \tag{9.3}$$

The carried traffic for the two-hop system is given as

$$C_2 = A(1 - p_2) = A[1 - \{1 - (1 - p)^2\}] = A(1 - p)^2 \tag{9.4}$$

From Eqs. (9.3) and (9.4)

$$\frac{C_2}{C_1} = 1 - p \tag{9.5}$$

The throughput of the two-hop system is thus always smaller than that of the single-hop system.

If we now make a reasonable assumption that the two-hop traffic consumes twice the network resources, then, for the same network resource used, we have

$$\frac{\text{carried two-hop traffic}}{\text{carried one-hop traffic}} = \frac{C_2}{2C_1} < \frac{1}{2} \tag{9.6}$$

Equation (9.6) can be interpreted as follows. For any finite p, the carried two-hop traffic is always less than half the one-hop traffic if the same network resource is deployed. This also implies that the price of the two-hop traffic should be more than twice that of the one-hop traffic if the former were to receive the same threshold of service as the single-hop customers. This result is independent of the topology of the network.

We can extend the result further and show that for any arbitrary network topology the compensatory price for a h-hop traffic must always be greater that h times the price of singe-hop traffic.

9.2 Pricing for Multi-hop Traffic

We now examine an equitable pricing plan for various classes of traffic in a multi-hop common user network where a class of traffic is defined by the number of hops between the source and the destination. As before, the network service provider can only charge the customer based on served traffic.

In developing the pricing model, we assume that the cost of transport is largely determined by the number of hops between the source and the destination. The cost of a two-hop transport for an identical bandwidth will be twice that of a single-hop transport. The cost of transport is determined as a function of the capacity or bandwidth of the link.

The circuit switched transport system is characterized by the Erlang B traffic model where a blocked traffic is lost and does not generate any revenue for the service provider. We then have p_1 (the blocking probability of single-hop traffic) $= p$. From the Erlang B formula [3]:

$$E_w(A_1) = \frac{\frac{A_1^w}{w!}}{\sum_{k=0}^{w} \frac{A_1^k}{k!}} = p \qquad (9.7)$$

The parameters of Eq. (9.7) are defined as follows: $w =$ Number of wavelengths or trunks; $A_1 =$ Incident traffic in Erlangs. The carried traffic $C_1 = (1 - p)A_1$ represents the traffic based on which the customer would be shared for the service.

9.2.1 Pricing for a Single-hop Traffic

For a single-hop network, there is only one link between the source and the destination. If the required quality of service is characterized by the probability of blocking p, then the throughput for the single-hop traffic is

$$E_w(A_1) = p \qquad (9.8)$$

where A_1 is the incident traffic at each node. In other words, if M_1 were the tariff associated with service, the revenue generated by the service provider can be given as

$$R_1 = M_1 A_1 (1 - p) \qquad (9.9)$$

9.2.2 Pricing for Two-hop Traffic

In [4], Kelly has shown that for large networks with multiple links in tandem, the end-to-end blocking probabilities can be evaluated by considering each link to be independent. This will allow us to make use of the Erlang B formula. Furthermore, reference [5] has proposed a method for computing the capacity-exhaustion probabilities for a large optical network with general holding-time distribution.

For the two-hop traffic, we have from Eq. (9.2)

$$E_w(A_2) = 1 - (1 - p)^{1/2} \tag{9.10}$$

The served two-hop traffic is $C_2 = (1 - p)A_2$.

Let M_2 be the tariff applicable to two-hop traffic, then

$$R_2 = M_2 A_2 (1 - p) = R_1 \tag{9.11}$$

Equation (9.11) represents the condition that the revenue from two-hop traffic equals revenue from one-hop traffic.

From Eq. (9.9) and Eq. (9.11), we have

$$\frac{M_2}{M_1} = \frac{A_1}{A_2} > 2 \tag{9.12}$$

It would be instructive to plot $\frac{M_2}{M_1}$ as a function of the required blocking probability such a plot would demonstrate that the compensatory tariff for the two-hop traffic would be always more than twice that for the single-hop traffic. Furthermore, the rate differential must increase as the minimum quality of service offered degrades. In other words, the lower the minimum quality the more expensive it is to serve higher classes of traffic.

We do note that, in addition to being compensatory to the service provider, the pricing scheme developed here is based on the optimum use of the transmission facilities. In other words, the two-hop pricing is also based on the fact that if the resources needed for two-hop traffic were deployed to carry single-hop traffic, the resulting revenue will be identical to what the two-hop traffic yielded.

9.2.3 Pricing for Multi-hop Traffic

The results of Sects. 9.2.1 and 9.2.2 can be generalized to include pricing for h-hop traffic. It can be shown that the h-hop traffic should be priced at a level higher than h times the price for single-hop traffic if the pricing system was based on the consumption of bandwidth while offering an identical grade of service to the user.

Traffic in a real network is, of course, from multiple classes; each class is defined by the parameter h which is the number of hops between the source and the destination. Pricing for simultaneous multiple classes of traffic is addressed in [1, 2].

9.3 Summary

This chapter has addressed how multiple classes of traffic affect the resource requirements of a network if the network was to provide a given threshold of blocking probability for all classes of traffic. Our analysis shows that traffic that carried over h links must have a tariff that is at least h-times the tariff for the single-hop traffic. The tariffs so computed will be compensatory to the service provider and equitable among the different classes of traffic.

9.4 Problems

9.1 A service provider has the following two options in choosing among the following two alternatives for transporting blocked and lost traffic from point A to point B:

- Option 1: A single line with a probability of blocking $= 0.1$
- Option 2: Two shorter lines in tandem; each with a blocking probability $= 0.1$. Assume that the blocking probabilities of the two lines in tandem are independent of each other.

Compute the fraction of the traffic that option 2 can carry relative to option 1. If the cost of each of the lines (both the shorter and the long lines were identical—generally, a fair assumption), compute the price difference between the two options, assuming that price was proportional to delivered traffic.

9.2 Explain the motivation for reducing the number of hops between the source and the destination for circuit switched traffic.

9.3 Even though the benefits of reducing the number of hops between the source and the destination are well understood, why do networks in general never have their nodes fully interconnected?

9.4 Can you translate the benefits of end-to-end communication over a single hop to transportation networks, e.g., air transport networks? Discuss the associated benefits and challenges.

References

1. Y. Qu, P.K. Verma, Notion of cost and quality in telecommunication networks: an abstract approach. IEE Proc. Commun. **152**(2), 167–171 (2005)
2. Q. Yingzhen, Enhancing the Traffic Carrying Capacity of Optical Networks, Ph.D. Dissertation, University of Oklahoma, Oklahoma, 2005
3. https://en.wikipedia.org/wiki/Erlang_(unit)
4. F.P. Kelly, Blocking probabilities in large circuit-switched networks. Adv. Appl. Probab. **18**, 473–505 (1986)
5. K.T. Nayak, K.N. Sivarajan, Routing and dimensioning in optical networks under traffic growth models: an asymptotic approach. IEEE J. Sel. Areas Commun. **21**(8), 1241–1253 (2003)

Chapter 10
Pricing of Packet Switched Services

10.1 Pricing Based on Bounded Delays

Delay is a key measure of performance in a packet switched system. Average delay has been widely used as an indicator of the quality of service in packet switched networks. Chapter 5, Sect. 5.3, has listed the components of delay. For a given network, with specified link transmission speeds, most of these delay elements, except the queuing delay, are constant or vary within narrow margins. However, since the telecommunication traffic is generally characterized by a Poisson process, the queuing delay can vary widely from instant to instant depending on the traffic intensity. Queuing delays can be reduced by increasing the speed of transmission links.

Chapter 7 has proposed the upper bound of delay as a key factor in characterizing packet switched networks. Traffic that exceeds the specified bound of delay is treated as lost traffic, does not constitute throughput, and the user does not pay for the traffic which has been delayed beyond the upper bound of the delay specified by the network provider. This delay is also termed the threshold delay. Indeed, it is possible to penalize the service provider further by increasing the service provider's penalty by a larger factor, say, 10 times the volume of traffic that exceeded the threshold delay.

10.1.1 Pricing for Single-hop Networks

Just as in the case of circuit switched networks, single-hop packet switched networks, shown in Fig. 7.2, serve packet switched traffic most efficiently.
From Eq. (7.1) we note that

© Springer Nature Switzerland AG 2020
P. Verma, F. Zhang, *The Economics of Telecommunication Services*, Textbooks in Telecommunication Engineering, https://doi.org/10.1007/978-3-030-33865-7_10

$$P\{W \le t\} = 1 - \frac{\lambda}{\mu C}e^{-(\mu C - \lambda)t} = 1 - \rho e^{-\mu C(1-\rho)t} \tag{10.1}$$

where P is the probability that a packet suffers a queuing delay less than or equal to t. λ and μ are the traffic parameters and C is the channel capacity. The throughput γ containing all packets with queuing delay less than or equal to t can be given by

$$\gamma = \lambda P \tag{10.2}$$

The throughput per unit of capacity deployed can be given as

$$\frac{\gamma}{C} = \frac{\lambda}{C}\left[1 - \frac{\lambda}{\mu C}e^{-(\mu C - \lambda)t}\right] \tag{10.3}$$

The optimum channel capacity C_{opt} for the given traffic parameters can be determined by putting:

$$\frac{\partial(\gamma/C)}{\partial C} = 0 \tag{10.4}$$

and showing that

$$\frac{\partial^2(\gamma/C)}{\partial C^2} < 0 \tag{10.5}$$

Equation (10.4) results in

$$\mu C_{opt} e^{(\mu C_{opt} - \lambda)t} = \lambda(2 + \mu C_{opt}t) \tag{10.6}$$

Again, C_{opt} is the optimum level of capacity needed for the system to deliver maximum throughput per unit of bandwidth (The second derivative as in Eq. (10.5) can be easily shown to be negative).

In developing a pricing model for the single-hop packet switched network, we treat the used capacity as the underlying cost and the throughput constituted by packets with queuing delay less than or equal to as the revenue. Under these assumptions, the cost for a unit of service delivered is equal to $\frac{C_{opt}}{\lambda P}$. A customer achieving a throughput γ will be charged an amount equal to

$$\frac{C_{opt}}{\lambda P}\gamma \tag{10.7}$$

A numerical example might clarify the approach adopted.
Let $\lambda = 2000$ packets per second, $1/\mu = 1$ bit, $t = 5$ ms.
Using Eq. (10.6), we can compute the optimal volume of capacity needed as

$$C_{opt} = 2490.3 \, bits/s.$$

The corresponding throughput γ can be evaluated from Eq. (10.3) as

$$\gamma_{max} = 2000 \left[1 - \frac{2000}{2490.3} e^{-(2490-2000)\dot{0}.005} \right] = 1861.4 \, packets/s.$$

This is the throughput the customer would achieve if the packets delayed more than 5 ms are to be discarded as not constituting throughput.

For a given amount of transmission resource and a specified intensity of traffic, the single-stage packet switched network serves that traffic most efficiently. If the same volume of traffic were routed through an intermediate server, queuing delays will be incurred at each server and, for a specified end-to-end threshold delay, the served traffic will be smaller. Correspondingly, if the served traffic and the end-to-end threshold delay were to remain constant, the transmission resource will have to increase at each server in order to deliver traffic at the receive end within the specified threshold of delay. The additional cost incurred by the service provider for delivering traffic over two hops will have to be factored in computing the charge for the two-hop traffic.

10.1.2 Pricing for Two-hop and Multi-hop Networks

Figure 7.2 shows a two-hop packet switched network and traffic delivered over two hops between the source and the destination. We have the probability P_2 for the end-to-end delay $\leq t$, from Eq. (7.11) given as

$$P_2(W \leq t) = \rho^2 \left\{ 1 - e^{-\mu C(1-\rho)t} [1 + \mu C(1 - \rho)t] \right\} \tag{10.8}$$

The throughput γ_2 for the two-hop traffic can be given as

$$\gamma_2 = \lambda P_2 \tag{10.9}$$

The subscripts used in P and γ, i.e., P_2 and γ_2, simply emphasize two hops between the source and the destination.

A comparison of Eqs. (10.1) and (10.8) representing the one-hop and two-hop traffic might be in order. We make the following observations:

- The delivered traffic for the two-hop system will be substantially lower compared to a one-hop system.
- If the two-hop network were to meet the latency requirement of a single-hop network, it would need two links, each with a capacity higher than in the single-hop network.

As shown in Chap. 7, computation of the price for the two-hop traffic will be based on the volume of traffic an identically resourced single-hop network could have accomplished and the revenue the single-hop network would have generated. The pricing will be, thus, based on the cost of lost opportunity as opposed to the consumption of resources.

The degradation of throughput of multi-hop traffic in a packet switched network can be addressed under similar assumptions. We simply note the following result for n-hop traffic [1]:

If we assume that each node of a n-hop VoIP network is an individual M/M/1 system and is characterized by the parameters μC and the threshold delay t, then the maximum throughput on an end-to-end basic is

$$\gamma_{n\ max} = \frac{\lambda_0^{n+2}}{(n-1)!}\left(\frac{t}{\mu C}\right)^n (\mu C - \lambda_0)e^{-(\mu C-\lambda_0)t} \tag{10.10}$$

This condition holds when the incident traffic λ_0 satisfies the following condition:

$$e^{(\mu C-\lambda_0)t} = \frac{\lambda_0 t[(\mu C-\lambda_0)t]^{n-1}}{(n+1)(n-1)!} + \sum_{k=1}^{n}\frac{[(\mu C-\lambda_0)t]^{n-k}}{(n-k)!} \tag{10.11}$$

The pricing structure for the multi-hop network can be developed in a fashion similar to the pricing structure of two-hop networks.

10.2   Pricing Based on Bounded Jitter

Jitter is a potential source of quality degradation in packet switched networks serving interactive communication. We now consider how jitter is affected by the number of hops a packet travels over several hops. Correspondingly, we also address how the network resources would have to be scaled up as the number of hops between the origin and destination of the network increases.

10.2.1   Pricing for Single-hop Networks

Jitter in a packet switched network is characterized by the variance of the delay or σ_D^2 and is given as [2]:

$$\sigma_D^2 = \frac{1}{\mu^2 C^2(1-\rho)^2} \tag{10.12}$$

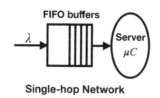

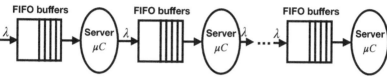

Fig. 10.1 Single-hop and multi-hop networks

Our assumption in developing Eq. (10.12) is that the packets follow the M/M/1 discipline and

- $1/\mu =$ mean packet length (bits),
- $C =$ speed of transmission (bits/s),
- $\rho = \lambda/\mu C$ where, $\lambda =$ rate of arrival of packets (no. of packets/s).

Figure 10.1 shows packet switched traffic served by a single, or multiple server in tandem. In evaluating the impact of multiple servers in tandem, we assume that the packets continue to follow the Poisson discipline for the arrival process at each hop.

10.2.2 Pricing for Multi-hop Networks

In order to simplify the analysis of the multi-hop network, we assume that each hop gets an identical volume of traffic with the same rate of arrival and mean packet length, and each server has the same server speed and that the arriving packets at each server are Poisson distributed. Although these assumptions are unrealistic in practice, the analysis will still offer an insight into the behavior of the end-to-end jitter associated with packets that transition through more than one hop.

We assume that the traffic at each hop is identical and there are n hops, it can be shown that the cumulative jitter is given by

$$\sigma_D^2 = \sum_{i=1}^{n} \sigma_{D_i}^2 = n\sigma_{D_i}^2 \tag{10.13}$$

The end-to-end jitter of a n-hop network can thus be shown to be

$$\sigma_D^2 = \frac{n}{\mu^2 C^2 (1-\rho)^2} \tag{10.14}$$

From Eq. (10.14), we can derive

$$C = \frac{1}{\mu}\left(\lambda + \sqrt{\frac{n}{\sigma_D^2}}\right) \tag{10.15}$$

Equation (10.15) can be, alternatively, presented as

$$\lambda = \mu C - \sqrt{\frac{n}{\sigma_D^2}} \tag{10.16}$$

Equation (10.16) presents the deterioration of the traffic carrying capacity of a multi-hop network in comparison to a single-hop network if the end-to-end jitter were to remain bounded to the same threshold as a single-hop network. Pricing for the multi-hop network has to be, therefore, increased to not only compensate for the additional transmission resource needed but based on what the total resource would have delivered if it were deployed to serve single-hop traffic and the revenue it would have generated.

10.3   Summary

This chapter has presented solutions relating the impact of bounding delays or jitter to a specified maximum threshold on resource requirement for a given volume of traffic and a specified network topology. Once the resource requirement is known, pricing is determined based on what level of revenue that resource would have generated if it were deployed in the most efficient manner.

Problems

10.1 Explain how you would use the learnings from this chapter in the design of networks from ground up.

10.2 One way to increase the efficient use of a network would be to offer preferential treatment to traffic with a lower number of hops between the source and the destination. Under heavy load situations, such a scheme might result in a drastic reduction in completing transfers that involve a large number of hops. How would you offer some level of throughput for multi-hop traffic?

References

1. P. Verma, L. Wang, *Voice over IP Networks* (Springer, New York, 2011)
2. M.H. Dahshan, P.K. Verma, Resource based pricing framework for integrated service networks. J. Netw. **2**, 36–45 (2007)

Chapter 11
Regulation

11.1 Need for Regulation

Regulatory oversights are common for markets where there is little or no competition. In contrast, in competitive markets, prices keep in control because of competition. For much of the twentieth century, the telecommunication network services market, globally, was deemed to be a natural monopoly. Markets that have a high barrier to entry are candidates for natural monopolies. Telecommunication network services market does have high barriers to entry. This is perhaps the most important reason why the telecommunication network services market was deemed to be a natural monopoly in most countries.

It is well known that a natural monopoly, in the absence of any public safeguards, would likely raise prices where it maximizes its profit at the expense of public welfare. Regulatory mechanisms were created to address this behavior of the telecommunication network services market.

In some cases, barriers to entry can be removed or minimized through regulatory intervention. One example of this is by requiring the dominant supplier of services to lease its services, facilities, equipment, or other assets to its emerging competitors at the elemental level and at prices that would make the emerging service providers competitive.

11.2 Forms of Regulatory Safeguards

For most of the twentieth century, in most countries, the natural monopoly problem of the telecommunication network services market was addressed by the government by owning the telecommunication network services market outright. The investment needed and the disposition of revenues were part of government's operations as in the other facets of the government's obligations to its people.

© Springer Nature Switzerland AG 2020
P. Verma, F. Zhang, *The Economics of Telecommunication Services*, Textbooks in Telecommunication Engineering, https://doi.org/10.1007/978-3-030-33865-7_11

In the USA, however, this market was owned by private business entities. This required the creation of a regulatory mechanism often involving jurisdictions that included the central government, the state governments, and the municipal governments.

In its most simplistic form, pricing of the public switched telecommunication network (PSTN) was parsed into two components: (a) a flat rate independent of usage and (b) usage. The latter is based on the number and duration of calls made, and their destinations.

Early regulatory mechanisms developed into two forms: (a) Rate of return regulation and (b) Price cap regulation.

11.2.1   Rate of Return Regulation

The rate of return regulation allows the telephone service provider to charge retail rates that were sufficient, in the aggregate, to cover its total expenses plus a reasonable rate of return on their investment. Of necessity, the expenses were based on historical costs, and historical lifetime of the assets the services provider would need to acquire, to offer services to customers. It is not difficult to see that both of these are subject to interpretations. Naturally, such differences would require addressing them through extensive negotiations or interventions by courts of law, as appropriate.

As one would anticipate, the rate of return regulation would result in the service provider "gold plating" its assets and prolonging the depreciation cycle of those assets as much as possible. The latter action would result in keeping the asset base high with a correspondingly high higher allowable revenue. In the Rate of Return regulation, there would be little incentive for the service provider to reduce the cost of its assets by, for example, taking advantage of new and emerging technologies and inducting them into its network. An alternative to the rate of return regulation was the price cap regulation discussed next.

11.2.2   Price Cap Regulation

The rate of return regulation tends to give the incumbent service providers incentives to "gold plate" their assets, that is, to spend more than is efficient or necessary simply to increase the rate, thus to increase profits. In the 1980s and 1990s, federal and state regulators sought to address these problems by adopting a price cap scheme for retail rate regulation of the incumbents. A price cap is set for a particular year using the rate of return regulation. In subsequent years, the price is determined by taking into account the following two factors. The first is driven by technological changes and other innovations resulting in industry-wide increases in efficiency; the second by changes in inflation and other factors related to economy. In contrast to the traditional rate of return regulation, incumbents are rewarded when they deploy more efficient equipment in the future.

The price cap, however, was a moving target. It reflected the anticipated business efficiencies based on cost trends of the components that constitute the network, inflation, the generally rising cost of labor, and other factors that would affect the underlying costs of the network.

It is not difficult to see that any part of a regulatory mechanism in a monopoly market would carry the risk of cross subsidization of one service by another. Disputes related to cross subsidization offer a fertile playground for regulators, service providers, and courts of law. The resolution of such disputes is generally long, often extending to months and years and, sometimes, decades.

With the evolution of the regulatory apparatus, first in 1984 by divesting the long-distance operations from the "local" calls, and then by the Telecommunications Act of 1996, and further deregulation especially with emergence of the widespread Internet, regulatory mechanisms have weakened considerably. This will be a continuing trend for the future. It is most likely, however, that regulatory oversight will continue to remain with the telecommunication network services for an indefinite period.

11.3   Emerging Regulatory Apparatus in the USA

The regulatory system in the USA has undergone massive changes in the past 35 years beginning with a transformational step in 1984 which broke the monopoly of AT&T in the long-distance telecommunication market. Regulatory changes are driven by the policy of the government and are expected to be in conformity with the constitution and legislation. Any discrepancy among the different arms of governance is interpreted by the courts of law and is binding among the concerned entities. For an interesting perspective on recent policy and legislative developments, we refer the readers to [1].

In principle, regulators act on behalf of the people and attempt to achieve the right balance between fairness across the spectrum of people in their constituency, while at the same time increasing the surplus of the society. Most often, regulators try to ensure that players in the market do not make excessive profit at the expense of the clients they serve.

An important aspect of regulation is, thus, ensuring that service providers in a regulated market make "reasonable" profit. While a regulator can delve into finer aspects of defining pricing for, say, different qualities of service in a multi-service market, such a step would be fraught with differences of opinion and the ensuing litigation. As a general rule, regulatory intervention in a non-commodity environment does not work too well.

The approach we adopt in this book is based on a segmented market with customers willing to pay different prices for different qualities of service. We address a regulator's concern by keeping the service provider's revenue and profit constant. This will allow the service provider to be market responsive while at the same time ensuring that prices to customers are kept at a reasonable level.

In this chapter, we address the regulatory framework under a circuit switched environment. The overall revenue of the service provider is kept constant while the service provider is allowed to offer differentiated pricing to its customers in a segmented market. We postulate that this is a valid scenario since different classes of customers may have the need for, and the resources to pay for, different classes of service in so far as the quality of service is concerned. In this chapter, we consider the Blocked and Lost call (or the circuit switched) network. A constant revenue model for packet switched network is considered in Chapter 16.

11.4 A Constant Revenue Model for Blocked and Lost Networks

In a blocked and lost network, which is essentially a circuit switched network, the quality of service is measured by the probability of blocking. A higher probability of blocking results in a lower quality of service and, of course, a lower throughput since a larger fraction of call attempts would be lost. This would result in a lower revenue for the service provider.

Literature on the subject is based on selective prioritization or wavelength reservation for different classes of service under varying incident traffic scenarios [2–6]. Note that wavelength reservation is equivalent to reserving a certain number of trunks for different classes of customers in the legacy network.

In developing a constant revenue model for the blocked and lost networks, we adopt an approach based on [7]. Our approach is based on the following: Under overload scenarios, a choice is offered to the incoming traffic. Customers requesting a higher grade of service are offered a higher price which is equivalent to a lower probability of blocking. The customer has, of course, the option to not choose the higher-price higher-grade service and, in that case, will be served at the lower price for a lower grade of service.

The pricing scheme developed will keep the overall revenue of the service provider constant mitigating the regulator's concern that the service provider might be engaged in an unfair business practice taking advantage of a situation when the demand exceeds the supply. The resources in the network, typified as the number of trunks, are kept constant. Since the resources in the network are at a constant level, the profit of the service provider is the same as before, mitigating the regulator's concern.

The validity of this approach is based on the fact that a class of customers might indeed prefer a lower price at a lower grade of service. Note that the differentiated pricing scheme is invoked only when the incoming traffic level exceeds a predefined value of incident traffic. For a lower volume of incident traffic, there is no need or motivation for instituting a differentiated pricing scheme. Any common user network can experience an inordinate level on demand by its customers. Should such a situation arise, the proposed scheme will offer a managed solution to serve its customers.

11.4.1 Mathematical Construct

The proposed model is shown in Fig. 11.1. For simplicity, the network is represented by a single link of N trunks or a DWDM facility with N channels. The assumption of a single link will simplify the approach but, more importantly, offer an insight into the operation of the scheme.

We assume that the network is designed to carry a level of traffic in Erlangs equal to A_n at a given probability of blocking equal to p_n. The price is maintained at a constant level P_0 as long as the traffic is less than A_n.

$A_n, p_n,$ and P_0 are the threshold parameters. Should incoming traffic level exceed A_n, a new probability of blocking p_b will come into play. As stated, the new pricing scheme will kick in at this point. Customers will be offered a choice as to whether they would like to maintain the probability of blocking p_n that existed prior to the network entering an overload situation at a higher price P_p or at the previous price P_0 with a higher probability of blocking.

- Case 1: **Incident Traffic A Lower than Threshold Traffic A_n—Prior to the Priority Scheme Kicking in**

  Figure 11.1 shows the configuration discussed below. We have, from the Erlang B formula,

$$p_b = E(A, N) = \frac{\frac{A^N}{N!}}{\sum_{i=0}^{N} \frac{A^i}{i!}} \tag{11.1}$$

The throughput can be given as,

$$S = (1 - p_b) \cdot A \ Erlangs \tag{11.2}$$

If the price charged to the customer is \$P/Erlang, the revenue R in dollars can be represented as,

$$R = S \cdot P \tag{11.3}$$

Fig. 11.1
Telecommunication link
$p_b < p_n$

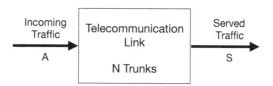

Note that for the case when the blocking probability reaches its threshold value p_n, the corresponding revenue is,

$$R_n = (1 - p_n) \cdot A_n \cdot P \qquad (11.4)$$

where

$$p_n = E(A_n, N) \qquad (11.5)$$

R_n is the maximum revenue the network service provider can collect from customers if the probability of blocking were to remain lower than the threshold probability of blocking p_n.

The incident traffic and the probability of blocking will have the characteristic shown in Fig. 11.2.

• Case 2: **Incident Traffic $A >$ Threshold Traffic A_n**

Figure 11.3 shows the configuration for the case when the incident traffic exceeds the threshold traffic.

In this case, priority is introduced and the service provider institutes two classes of service.

The low priority service will receive a higher probability of blocking while the high priority service will receive the threshold probability of blocking which is p_n. A certain number of trunks will be reserved for the higher priority service and the rest will be deployed for serving the low priority customers. As before, let, $N =$ Total number of trunks, $A =$ Incident traffic in Erlangs. We split N into two distinct groups,

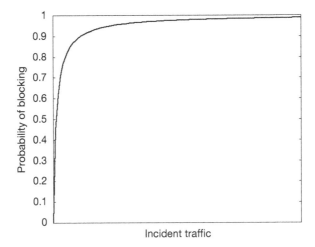

Fig. 11.2 Probability of blocking as a function of incident traffic

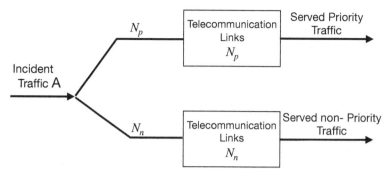

Fig. 11.3 Incident traffic $A >$ threshold traffic A_n

$$N = N_p + N_n \tag{11.6}$$

where N_p serves the higher priority customers and N_n serves the low priority customers.

Let q be the fraction of traffic choosing priority at a price P_p. We compute N_p from the Erlang B formula,

$$p_n = E(qA, N_p) \tag{11.7}$$

The throughput of the priority traffic is given as,

$$S_p = (1 - p_n) \cdot q \cdot A \tag{11.8}$$

We have N_n trunks serving non-priority traffic of $(1 - q)A$ Erlangs. The probability of blocking of non-priority traffic is given from the Erlang B formula:

$$p_{np} = E((1 - q)A, N_n) \tag{11.9}$$

To Keep the revenue constant at R_n, different prices need to be charged to priority and non-priority traffic P_p and P_{np}, respectively. These prices are calculated dynamically from the following relations:

$$P_p = \frac{N_p}{N} \cdot R_n \cdot \frac{1}{qA \cdot (1 - p_n)} \tag{11.10}$$

$$P_{np} = \frac{N_n}{N} \cdot R_n \cdot \frac{1}{(1 - q)A \cdot (1 - p_{np})} \tag{11.11}$$

Note that in Eqs. (11.10) and (11.11), the price of high priority and low priority traffic is determined in a way such that the overall revenue of the network is kept constant at a value equal to the threshold revenue that would exist if the network worked at its full capacity at the threshold probability of blocking. The overall revenue collected from each group is proportional to the resources deployed for the respective groups.

The proposed scheme thus allows the overall revenue of the service provider to remain constant, the high priority customers receiving the probability of blocking equal to the threshold value and the low priority customers receiving a higher probability of blocking depending on the number of trunks allocated to them and the intensity of low priority traffic.

We do note that the proposed scheme might work only within limited ranges of q and the total traffic intensity and resources. Should this be the case not, the network resources would warrant adjustment as needed.

11.5 Summary

This chapter has introduced legacy regulations for public switched telecommunication networks and discussed a constant revenue model for a circuit switched network.

Problems

11.1 Characterize the difference between pricing schemes based on: (1) rate of return and (2) price cap. Which one of these is conducive to the adoption of new technology into the network? Why?

11.2 A service provider in a monopoly environment is regulated on a rate-of-return basis. Explain why the provider would have the incentive to invest in equipment with a long depreciation cycle.

11.3 A service provider is attempting to subsidize a service A by another service B in a competitive market. Explain the kind of risk it faces from a new entrant.

11.4 Explain the rationale for the constant revenue model from a regulator's perspective.

11.5 The constant revenue model might not work well when the incident traffic rises arbitrarily high. If this situation arises frequently, what recommendation would you offer to the service provider?

References

1. J.E. Nuechterlein, P.J. Weiser, *Digital Crossroads*, 2nd edn. (MIT Press, Cambridge, 2013)
2. Y. Qu, P.K. Verma, Limits on the traffic carrying capacity of optical networks with an arbitrary topology. IEEE Commun. Lett. **8**(10), 641–643 (2004)
3. Y. Qu, P.K. Verma, J.Y. Cheung, Wavelength reservation and congestion control in optical networks, in *Proceedings of the IASTED International Conference on Optical Communication Systems and Networks* (2004), pp.790–795
4. Y. Qu, P.K. Verma, J.Y. Cheung, Enhancing the carrying capacity of a DWDM network, in *Proceedings of the 2004 International Conference on Parallel Processing Workshops, Montreal, Canada, Aug. 15–18* (2004)
5. C. Lemieux, Theory of flow control in shared networks and its applications in the Canadian telephone network. IEEE Trans. Commun. **COM-29**, 399–413 (1981)
6. S. Yaipairoj, F.C. Harmantzis, Dynamic pricing with "alternatives" for mobile networks, in *Wireless Communication and Networking Conference, 2004. WCNC, 21–25 March 2004*, vol. 2. (IEEE, Piscataway, 2004), pp. 671–676
7. H. Sukasdadi, P. Verma, A constant revenue model in telecommunication networks, in *Proceedings of the International Conference on Networking, Morne, Mauritius, Apr. 23–29* (2006)

Chapter 12
Net Neutrality

12.1 Introduction

The term, "net neutrality" is a relatively recent term in the parlance of telecommunication networks. It was first proposed by Columbia Law School professor, T. Wu, in 2003 [1].

Prior to the advent of broadband networks toward the end of the twentieth century, the legacy public switched telecommunications network (PSTN) offered a mechanism that allowed analog voice communication between any two endpoints of the network. When the need for data to be transported between any two endpoints first arose, the PSTN could meet that need expeditiously. Existing protocols and network procedures were used for the call set up. Once an end-to-end physical channel was set up, the use of a modem allowed the analog 3 kHz channel to become a transport mechanism to carry digital signals from a source to the intended destination.

The PSTN could be characterized by three attributes: ubiquity, a fixed bandwidth, and a level of performance consistent with the transport of one voice channel [2]. Ubiquity was a definite advantage resulting in universal accessibility. The fixed bandwidth was a definite disadvantage because of the growing needs of machines to communicate at ever higher speeds.

The PSTN provider had no control over the circuit which will be selected by the dial-up network. It had thus no control over the quality of the circuit or the bandwidth which was, in this case, fixed. Once an end-to-end circuit was selected, the service provider could not exercise any control over the duration of the call or affect the throughput or the quality of the circuit. The data transport architecture with PSTN as the core network is shown in Fig. 12.1.This simple model of networking was both universal and, to the extent that the service provider could not control the speed or the latency, it was also non-discriminatory.

The Integrated Services Digital Network (ISDN) architecture, proposed and implemented on a limited basis in the mid-1980s, created a digital network that

© Springer Nature Switzerland AG 2020

P. Verma, F. Zhang, *The Economics of Telecommunication Services*, Textbooks in Telecommunication Engineering, https://doi.org/10.1007/978-3-030-33865-7_12

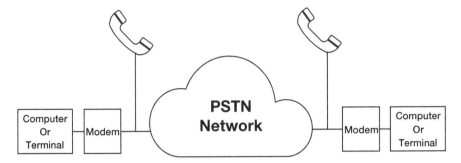

Fig. 12.1 PSTN-based data communication

would allow end-to-end communication at 64 kbps. The access network offered two 64 kbps data channels. The ISDN, however, was a short-lived phenomenon as the demand for networking at higher speeds continued to rise beyond what the PSTN-ISDN service providers could have anticipated. This emerging need led to the core network adopting a packet switching architecture. Packet switching accommodates a variety of applications economically and would, for this reason, become the transport vehicle for the Internet.

The Internet with a packet switching architecture at its core accommodates a vast range of speeds and offers an economy of scale directly attributable to removing the constraints associated with channels of fixed bandwidth. A packet switching architecture of the core network and broadband access speeds resulted in new applications over the network, not envisaged earlier. Simultaneously, the wave of deregulation and a multiplicity of operators in the networking business presented regulatory challenges not foreseen earlier. These are discussed in the following section.

12.2   Stakeholders in the Broadband Access Era

The new world of broadband access would have more stakeholders than the previous telecommunications networking world where, in addition to the user, there were only two stakeholders: the service provider and the content provider. The regulatory instruments in this legacy environment put a clear functional separation between the network services provider and the content providers. The content owners could not offer services, and the service providers could not possess and distribute contents or information.

The new world of broadband access has four major stakeholders:

- The broadband access provider
- The Internet service provider
- The content provider
- The user

Additionally, any of the stakeholders could adopt multiple roles. For example, the Internet Service Provider could also own the broadband access facilities and could own the content as well. Regulatory apparatus in the USA and, possibly, in most other countries could, of course, regulate the functions of each of these stakeholders and prevent them from transgressing into the territories allocated to other stakeholders. Such a rigid separation of the functions allocated to different providers is, in theory, possible but it would increase the cost of doing business which will eventually be passed on to the end user, as increased charges. Furthermore, it would likely degrade performance at the end-user level. The performance degradation is attributable to the fact that whenever one administrative domain formally interfaces with another, there is usually protocol termination accompanied by the generation of another protocol and remote loop back capabilities so that each provider in the chain could validate the functionality within its control. The increased cost and performance penalty are again borne by the end user.

Another issue with the functional partition of an end-to-end process is the likely lack of innovation that could result if the providers were forced to operate in narrowly defined functional domains. And lack of innovation would affect not only the competitiveness of the telecommunication industry but the competitiveness of a nation as whole since telecommunication is such an integral part of all businesses. On the other hand, allowing a major supplier the ability to control the entire chain of functionalities between two or more end points could, possibly, result in unfair dominance of one, thus stifling competition. The net neutrality proponents are trying to avert the possibility that a dominant supplier could resort to predatory practices.

12.3 Net Neutrality

Simply put, net neutrality is the principle that all Internet traffic should be treated equally. If there is no functional separation of the end-to-end operational space among the broadband access provider, the Internet service provider, and the content provider, then, for example, an Internet service provider could, possibly, accord preferential treatment to its own content over a competitor's content. One example of preferential treatment by a network operator is selectively controlling the speed of a competitor's traffic over the network while according no control to the flow of its own content. This would create an inequity in the marketplace that would negatively affect an end user by compromising his or her flexibility to download content from a multiplicity of content owners. Furthermore, a content (only) provider would be always suspicious of the fact that the network services provider will choke or otherwise reduce the performance expected by the content provider's customers to preferentially advantage competing content owned by the service provider.

On the other hand, a broadband access provider would argue that a content provider would place an undue burden on its access network because the content provider would push an extremely high volume of data on the network compared to an end user who would utilize the access network at a considerably lower level.

The inequity between the high- and low-level users is a result of the fact that, by convention and in practice, the broadband access providers charge their customers based on the speed they offer to the customer, and not by the volume of data a customer pours into the network. Viewed in this fashion, net neutrality can be deemed to be a form of price regulation, which is what the wave of deregulation intended to do away with.

Given the inequity of usage between the light and the heavy users of the access network, the broadband access providers contend that they should be allowed to preferentially choke streaming traffic from the content providers in order to maintain the integrity of the network under heavy loads. This is where the crux of the net neutrality issue lies. A neutral net will not discriminate between traffic generated by the heavy and the light users or, indeed, traffic originated by or meant for any pair of users, including traffic requiring low latency or urgency.

Mistakenly, net neutrality is sometimes confused with the absence of digital divide. Digital divide is disparately lower availability of computing and/or Internet access resources to less affluent sectors of the population. It is generally believed that a relatively larger proportion of computing and communication resources are available to a narrow sector of the society. Since these resources are crucial to empowering individuals in the modern era, digital divide will further disadvantage those who already are disadvantaged. Perhaps the term "neutrality" connotes absence of undue favor to a sector that is already dominant. Viewed from this perspective, neutrality will connote leveling of the playing field between the affluent and less affluent members of the population. Those who share this connotation will believe that net neutrality will lead to a more egalitarian society.

12.4 The Sustainability of Net Neutrality

A neutral net will become a monolithic behemoth expected to treat all traffic equally, independent of the volume and the underlying characteristic of the traffic. For example, an interactive traffic (such as a phone conversation) requires a low latency in the tens of milliseconds region and is characterized by low bandwidth. On the other hand, television broadcast demands bandwidth in the megabits per second region but can withstand a second or more of delay without affecting the end user.

Increasingly, the Internet is being used to convey information that requires ultra-fast treatment such as those required by emergency service providers that deal with life-threatening situations. A neutral net will have no ability to discriminate between the traffic that must be immediately delivered, such as traffic that relates to life-threatening situations, and traffic that carries, say, routine broadcast material. This is not desirable from a societal good perspective. A totally neutral common user network is, therefore, an unsustainable solution over the long term.

12.5 Net Neutrality vs. Transportation Neutrality

It has been well known to engineers and scientists over the past several decades that transportation networks and telecommunication networks share features that make them similar from functional, performance, and investment standpoints. For starters, their incremental costs for serving a unit payload are nearly zero or, in the parlance of economists, they have zero marginal cost. Both these industries are capital intensive, meaning that their initial capital outlay is high, while the recovery of investment takes place over a prolonged period. Moreover, while a change in demand for both happens incrementally, capacity enhancements can occur only in big chunks.

The behavior of both these networks under excess load beyond their capacity is also nearly identical. For example, while any product or service offering organization will crave to have an excess of demand requesting their product or service, such a situation would be the worst nightmare for a communication or transportation network provider because the served traffic will deteriorate rapidly to levels far below the designed capacity if the incident load of the network far exceeds the designed capacity of the network. Furthermore, the capacity of the network cannot be added to in small quantities, or in an incremental manner.

Given this level of similarity, one would anticipate that the providers of communication and transportation networks, or indeed the providers of any service offered off a common user network, would require an identical approach in so far as the use of regulatory tools in managing their behavior for the public good is concerned. This has not been the case, however, and any similarity between the regulatory tools that address nearly identical issues in the two sectors of communication and transportation is noticeably absent.

The proponents of the now hotly debated net neutrality should ask themselves the following questions: Shouldn't vehicular traffic of all sizes, shapes, and weights from a two-wheeler carrying a single passenger to a 5-ton 18-wheeler pay the same toll? Should any restrictions on traffic lanes, especially those designated for heavy vehicles or high occupancy vehicles, be immediately abandoned in the interest of offering a neutral approach to all? In the same vein, should the toll on the toll roads be the same independent of vehicular size and weight even though they consume the capacity of the network to vastly different extents? Most importantly, should the approach toward regulating the investment in the transportation network be as capricious as it is in the world of telecommunications?

A key element in controlling the behavior of any of the stakeholders, or combinations thereof, in the broadband era will be transparency of the behavior of each of them toward the others. This will require each stakeholder to fully disclose the control, if any, it exercises on the traffic. As long as all traffic is handled identically without regard to the origin and destination, the network can be considered neutral. The characteristics of traffic can, however, still determine the level of control it might be subject to. This, combined with competition in the marketplace, will likely result in all the benefits of neutrality.

12.6   A Constant Revenue Model for Net Neutrality

Broadband service providers, like at&t, Verizon, and Comcast, among others, view net neutrality as being unfair to: (a) broadband service providers themselves and (b) light network users (compared to heavy users subject to the same tariff). The broadband service providers argue it is the service providers who have put their resources which they have to maintain and upgrade for their customers, and if the light users subsidize heavy users, the network will become resource constrained to the very detriment of content providers who are promoting net neutrality. Lack of additional sources of revenue will act as a disincentive for the broadband service provider to upgrade their infrastructure which, in turn, will affect the service provider's plans of increasing capacities. Furthermore, it is estimated that 80% of Internet traffic is caused by 5% of the user population. In order to keep network traffic flowing for all consumers, broadband service providers argue it is reasonable for them to use network traffic management practices.

This book adopts a different approach to addressing issues associated with net neutrality. We assume that all users are responsible for the cost of providing and provisioning the network, and we model this cost sharing problem as a cooperative game. The cost share associated with each user corresponds to its Shapley Value [3] of the corresponding cooperative game. We consider that the total cost of a coalition of users as the network resource required to maintain the desired QoS for the traffic in the coalition. Inter-user compensations are established based on the difference between their cost share, and the actual price they pay by way of access fees to the broadband service provider. The constant revenue model proposed in this book utilizes game theory techniques which are presented in the Chap. 18.

12.7   Summary

This chapter has identified the major stakeholders and their business interests in the net neutrality debate. It concludes that net neutrality as understood and proposed today is likely unsustainable over the long haul.

Problems

12.1 Name the principal stakeholders in the net neutrality debate. Why are the service providers and the content providers on the opposite sides of the debate on net neutrality?

12.2 One way to make the net neutral and reduce the power of the dominant player will be to separate the stakeholders through structural separation where the operational space of each stakeholder is rigidly defined. Comment on the merit of this approach. How would the net neutrality debate further intensify or weaken as competition among providers of the access platform increases? Why?

12.3 How does the net neutrality debate relate to the network offering mission critical applications?

References

1. T. Wu, Network neutrality, broadband discrimination. J. Telecommun. High Technol. Law **2**, 141 (2003). https://doi.org/10.2139/ssrn.388863.SSRN388863
2. P. Verma, The elusive goal of net neutrality. Int. J. Crit. Infrastruct. Prot. **4**(3+4) (2011). https://doi.org/10.1016/j.ijcip
3. A.E. Roth, R.E. Verrecchia, The Shapley value as applied to cost allocation: a reinterpretation. J. Account. Res. **17**(1), 295–303 (1979)

Chapter 13
Game Theory and Its Applications to Communication Networks

13.1 A Brief Introduction to Game Theory

A game consists of a principal (e.g., the network service provider) and a finite set of players (e.g., network users) $N = \{1, 2, \ldots, n\}$. The network service provider supports $M = \{1, 2, \ldots, m\}$ different QoS classes. Each player will then choose a strategy $x_i = \{x_{i1}, x_{i2}, \ldots, x_{im}\}$ with the objective of maximizing its payoff function u_i, where x_{ij} is the amount of traffic from service j that user i consumes. The following terminology applies to the classes of games we study:

- Player i's strategy, $i \in N$, is a M-dimensional vector x_i;
- A player's strategy space $X_i \subseteq R^M$ is the set of strategies available to user i, thus $x_i \in X_i$;
- A joint strategy x is the vector containing the strategies of all players: $x = \{x_1, x_2, \ldots, x_n\}$;
- The joint strategy space X is defined as the Cartesian product of the strategy spaces of all players: $X = \times_{i \in N} X_i$;
- Each player's payoff is a scalar-valued function of the joint strategy and we denote this function by $u_i(x) : X \to R$.

Games can be differentiated into non-cooperative games and cooperative games. In a non-cooperative game, each player chooses his or her strategy independently while in a cooperative game, players are able to form binding commitments and communications are always assumed to be allowed among players.

13.1.1 Non-cooperative Games

Game theory attempts to predict the outcome of such a game or, when this is not feasible, properties of the predicted outcome, such as its existence and uniqueness.

© Springer Nature Switzerland AG 2020
P. Verma, F. Zhang, *The Economics of Telecommunication Services*, Textbooks in Telecommunication Engineering, https://doi.org/10.1007/978-3-030-33865-7_13

This leads to the important definition of the Nash Equilibrium in a non-cooperative game, a joint strategy where no player can increase his or her payoff by unilaterally changing his or her strategy.

Definition 13.1 (Nash Equilibrium) Strategy $x \in X$ is a Nash equilibrium if $u_i(x) \geq u_i(x_i^*, x_{-i}), \forall x_i^* \in X_i, \forall i \in N$, where x_{-i} represents all components of vector x except its ith component.

The Nash equilibrium predicts the outcome of the game. Generally speaking, if all players predict that a Nash equilibrium will occur, then no player has an incentive to choose a different strategy [1]. In general, the uniqueness or even the existence of a Nash equilibrium is not guaranteed; neither is convergence to an equilibrium when the equilibrium exists.

A direct interpretation of Definition 13.1 is that the Nash equilibrium is a mutual best response from each player to the other players' strategies. In order to formally state this result, we first define the best reply mapping [2]:

Definition 13.2 (Best Reply Mapping) The best reply mapping for player i is a point to set mapping that associates each joint strategy $x \in X$ with a subset of X_i according to the following rule: $\psi_i(x) = \arg\max_{x_i^* \in X_i} u_i(x_i^*, x_{-i})$. The best reply mapping for the game is then defined as $\psi(x) = \times_{i \in N} \psi_i(x)$.

It is sometimes convenient to make use of the following alternate definition of Nash equilibrium [2].

Definition 13.3 (Nash Equilibrium) Strategy x is a Nash equilibrium if and only if $x \in \psi(x)$.

We emphasize that the idea of the Nash equilibrium as a consistent predictor of the outcome of the game does not necessarily require perfect knowledge on the part of the players regarding other players' payoff functions. Even without this knowledge, in a quasi-static environment, players may converge to an equilibrium through a learning process.

An equilibrium is expected to be efficient. Pareto optimality is used to determine the efficiency of Nash equilibrium. In the case of multiple equilibria, should one of these be Pareto optimal, it is considered to be better than others. Pareto Optimality is defined bellow.

Definition 13.4 (Pareto Optimality) A strategy x is Pareto optimal if there does not exist $x' \in X$ such that:

1. $u_i(x') \geq u_i(x), \forall i \in N$ and
2. $u_i(x') > u_i(x)$ for at least one $i \in N$.

Table 13.1 The original
Prisoner's Dilemma

		Colin	
--------		-------	-------
		A	B
Rose A		(0,0)	(−2,1)
B		(1,−2)	(−1,−1)

A is "don't confess," B is "confess"

13.1.2   Cooperative Games

In 1950, Melvin Dresher and Merrill Flood at the RAND Corporation devised a game (see Table 13.1) to illustrate that a non-cooperative game could have an equilibrium outcome which is unique. However, it is not Pareto optimal [3].

- If one of them confesses and the other does not, the confessor will get a reward (payoff + 1) and his partner will get a heavy sentence (payoff − 2);
- If both confess, each will get a light sentence (payoff − 1);
- If neither confesses, both will go free (payoff 0).

In the years since 1950 this game has become known as the Prisoner's Dilemma. Strategy B is dominant for both players, leading to the unique equilibrium at BB. However, this equilibrium is non-Pareto-optimal, since both players would do better at AA.

Instead of choosing their strategies independently, when we assume that a player can communicate and form a coalition (e.g., both promise to play strategy A), this is a cooperative game. Unlike the game we describe in Table 13.1 with only two players, in our modern connected world, most economic, social, and biology games involve more than two players. The questions are:

- Which coalitions should form?
- How should a coalition divide its winnings among its members?

Before looking into these questions, we first define characteristic functions.

Definition 13.5 (Characteristic Function) A game in characteristic function form is a set of N players, together with a function v which for any subset $S \subseteq N$ gives a number $v(S)$.

Players in the subset S, if they formed a coalition, could win the amount $v(S)$. The function v represents the characteristic function of the game. To calculate $v(S)$, assume that the coalition S forms and then plays optimally against an opposing coalition $N - S$.

There is an important relation among the values of different coalitions which holds for games in characteristic function form: superadditive.[1]

Definition 13.6 (Superadditive) A game (N, v) in characteristic function form is superadditive if $v(S \cup T) \geq v(S) + v(T)$ for any two disjoint coalitions S and T.

If two coalitions S and T, with no common members, decide to join together to form $S \cup T$, Definition 13.6 says that they can always assure themselves of at least $v(S) + v(T)$ because they can simply continue to do what they would do if they had not joined. And, they may often be able to do better than this by coordinating their actions.

From Definition 13.6, we find that it is in all players' interest to form a coalition N and get $v(N)$. Instead of asking about the possible results of actual coalitional behavior, we consider one class of cooperative games and ask if there might be a single payoff $|N|$-dimensional vector which could represent a fair distribution of payoffs to all players. This payoff vector might not arise from the competitive behavior of coalitions, but it would be the payoff vector an outside arbitrator might impose, taking into account the relative strengths of the various coalitions. For example, in Neumann and Morgenstern's Divide the Dollar game, three players will be given a dollar if they can decide how to divide the dollar among themselves by majority vote. We can see that a likely outcome might be that one of the three two-person coalitions would form and divide the dollar equally between its two members. As a result, we get one of the three payoff vectors $(\frac{1}{2}, \frac{1}{2}, 0)$, $(\frac{1}{2}, 0, \frac{1}{2})$, or $(0, \frac{1}{2}, \frac{1}{2})$. An outside arbitrator considers the symmetry of this game and decides the fair division is certainly $(\frac{1}{3}, \frac{1}{3}, \frac{1}{3})$. This is the case in cost sharing of multi-service communication networks. We do not prefer one coalition to another, but would like a fair distribution of cost among different classes of service.

In 1953, Lloyd Shapley gave a general answer to this fair division question and it has come to be known as the Shapley Value of a cooperative game in characteristic function form. We first look at three definitions which capture a fair distribution of payoffs [3]. Here, we use payoff vector $\varphi = (\varphi_1, \varphi_2, \ldots, \varphi_n)$ to denote the fair payoff to each player.

Definition 13.7 (Efficiency) The total gain is distributed: $\sum_{i \in N} \varphi_i = v(N)$ and therefore payoff allocation φ is Pareto optimal.

Definition 13.8 (Symmetry) φ should depend only on v and should respect any symmetries in v. That is, if plays i and j have symmetric roles in v, then $\varphi_i = \varphi_j$.

Definition 13.9 (Zero Player) If $v(S) = v(S - i)$ for all coalitions $S \subseteq N$, that is, if player i is a dummy who adds no value to any coalition, then $\varphi_i = 0$. Furthermore, adding a dummy player to a game does not change the value of φ_j for any other players j in the game.

Definition 13.10 (Additivity) $\varphi[v + w] = \varphi[v] + \varphi[w]$

---

[1] This holds for all games in characteristic function form which rise from games in normal form.

Table 13.2 Cost sharing problem using the Shapley Value

Order	Incremental cost		
	A	B	C
A, B, C	$v(A) - \phi = 1$	$v(A, B) - v(A) = 1$	$v(A, B, C) - v(A, B) = 1$
A, C, B	$v(A) - \phi = 1$	$v(A, B, C) - v(A, C) = 0$	$v(A, C) - v(A) = 2$
B, A, C	$v(A.B) - v(B) = 0$	$v(B) - \phi = 2$	$v(A, B, C) - v(A, B) = 1$
B, C, A	$v(A, B, C) - v(A, B) = 0$	$v(B) - \phi = 2$	$v(B, C) - v(B) = 1$
C, A, B	$v(A, C) - v(C) = 0$	$v(A, B, C) - v(A, C) = 0$	$v(C) - \phi = 3$
C, B, A	$v(A, B, C) - v(B, C) = 0$	$v(B, C) - v(C) = 0$	$v(C) - \phi = 3$
Total	2	5	11

The above definition is about the sum of two games. Suppose that (N, v) and (N, w) are two games with the same player set N. Then we can define the game $v + w$ by defining $(v + w)(S) = v(S) + w(S)$ for all coalitions S. Now we have three games under consideration and use $\varphi[v]$, $\varphi[w]$, and $\varphi[v + w]$ to denote payoff vector for each game. It means that if it is fair for player i to get $\varphi_i[v]$ in v and $\varphi_i[w]$ in w, it would seem fair to get the sum of these two payoffs in the game $v + w$. For example, the cost sharing of communication network with fixed cost and operation cost is the cost sharing of the fixed cost plus the cost sharing of the operation cost.

Shapley has proved that there is one and only one method of assigning payoff vector φ for a game (N, v) which satisfies all above definitions.

Shapley Value: The Shapley Value of each player i in the cooperative game has the following expression:

$$\varphi_i = \sum_{S \subseteq N \setminus \{i\}} \frac{|S|!(|N| - |S| - 1)!}{|N|!} [v(S \cup \{i\}) - v(S)] \tag{13.1}$$

The meaning behind the Shapley Value is that each player's payoff depends on the incremental cost for which he or she is responsible when provision of the services accumulates in random order. Let us illustrate it with an example.

Suppose there are three airplanes A, B, C sharing a runway. Airplane A requires 1 km to land, Airplane B requires 2 km, and Airplane C requires 3 km. When a runway of 3 km is built, how much should each airplane pay? Before we applying Eq. (13.1), we look at their incremental cost in the six possible orders in Table 13.2 (cost is measured in unites per kilometer):

Therefore, based on the Shapley Value Eq. (13.1), each airplane should be responsible for $(\frac{2}{6}, \frac{5}{6}, \frac{11}{6})$, respectively. Let us calculate this problem by treating it as sum of three games. The first kilometer is shared by all airplanes and so its cost should be $(\frac{1}{3}, \frac{1}{3}, \frac{1}{3})$; the second kilometer is shared by airplane B and C and their cost should be $(0, \frac{1}{2}, \frac{1}{2})$; the last kilometer is used only by airplane C and the allocated cost should be $(0, 0, 1)$. Based on the Definition 13.10, the cost sharing of this runway is the sum of above vector and the result is $(\frac{2}{6}, \frac{5}{6}, \frac{11}{6})$.

13.2 Game Theory in Communication Networks

Although the original applications of game theory tended to be in the field of economics [14], over the years its usefulness has been recognized in other disciplines such as biology, political science, and philosophy. Recently, game theory has been applied to numerous networking problems.

- **Spectrum Management**: In a communication system where multiple users share a common frequency bank such as cognitive radio, each user's performance, measured by a Shannon utility function, depends on not only the power allocation (across spectrum) of its own, but also those of other users in the system. This spectrum management problem can be formulated either as a non-cooperative game [4, 7, 8] or as a cooperative utility maximization problem [9, 10]. In [4], each user is given an initial budget to purchase its own transmit power spectra (taking others as given) in order to maximizing its Shannon utility or payoff function, which includes the effects of interference. It has proved that an equilibrium always exists for a discrete version of the problem, and, under a weak-interference condition or the Frequency Division Multiple Access (FDMA) policy, the equilibrium can be computed in polynomial time.
- **Flow Control**: Papers [5, 6] study the issue of flow control using a game theoretic framework. Reference [6] defines each flow's utility function as its transmission rate and the QoS it receives and model this interdependence under a game theoretic framework. Instead of using edge routers to shape incoming flows, this paper has assumed that flows in the network are responsible and has proved there exists a Pareto optimal equilibrium. In [6], each user's objective is to maximize the average throughput subject to an upper bound on average delay. Since users share a network of quasi-reversible queues, each user's strategy affects all other users' performance and the authors have determined the existence of an equilibrium for such a system.
- **Congestion Control**: Reference [11] considers the problem of whether a given switch service rule will lead to an operating point that is fair and efficient or not using game theory. In [11], users' payoff function is defined as the amount of traffic and received QoS provided by a switch and users are selfish to maximize their payoff by varying the level of traffic.
- **Routing**: References [12, 13] have investigated routing issues in communication networks using the game theory model. Reference [12] has studied the existence of routing strategies of the network manager that drives the system to an optimum operating point. Reference [13] has investigated how to partition the traffic from a number of users among a number of parallel links to maximize their payoff function, which is defined as a measure of performance and satisfaction.
- **Pricing**: Several multi-service network pricing schemes using game theoretic models are described in Sect. 8.5.3.

Game theory provides a powerful tool for studying the performance and QoS issues in communication networks. In this book, we study the pricing issue of multiservice networks under a game theoretic framework.

13.3 Summary

This chapter has introduced background information on game theory: The Nash Equilibrium in non-cooperative games and the Shapley Value in cooperative games. Additionally, it has reviewed the applications of game theory in communication networks, including spectrum management, flow control, congestion control, and routing.

In the next chapter, we will introduce network and pricing models used in this book.

Problems

13.1 Define a non-cooperative game and a cooperative game.

13.2 Define Nash equilibrium.

13.3 Define Pareto Optimality.

13.4 Find the Shapley Value for each player in the following cooperative game: There are three players $\{1, 2, 3\}$. Player 1 has a right-hand glove, player 2 has a right-hand glove, and player 3 has a left-hand glove. The goal is to form pairs. The value function for this game is: $v(S) = 1$ if $s \in \{\{1, 3\}, \{2, 3\}, \{1, 2, 3\}\}$; 0 otherwise.

References

1. D. Fudenberg, J. Tirole, *Game Theory* (MIT Press, Cambridge, 1993)
2. J.W. Friedman, *Game Theory with Applications to Economics*, 2nd edn. (Oxford University, Oxford, 1990)
3. P.D. Straffin, *Game Theory and Strategy* (Mathematical Association of America, Providence, 1993)
4. Y. Ye, Competitive communication spectrum economy and equilibrium, in *Working Paper* (2007)
5. Y.A. Korilis, A.A. Lazar, On the existence of Equilibria in noncooperative optimal flow control. J. Assoc. Comput. Mach. **42**(3), 584–613 (1995)
6. F. Zhang, P.K. Verma, A two-step quality of service provisioning in multi-class networks, in *Proceedings of the18th International Conference on Telecommunications (ICT 2011), Ayia Napa, Cyprus* (2011)
7. Z.Q. Luo, J.S. Pang, Analysis of iterative water-filling algorithm for multi-user power control in digital subscriber lines, in *Special issue of EURASIP Journal on Applied Signal Processing*

on Advanced Signal Processing Techniques for Digital Subscriber Lines, vol. 2006 (2006), Article ID 24012

8. N. Yamashita, Z.Q. Luo, A nonlinear complementarity approach to multi-user power control for digital subscriber lines. Optim. Methods Softw. **19**, 633–652 (2004)

9. W. Yu, R. Lui, R. Cendrillon, Dual optimization methods for multi-user orthogonal frequency division multiplex systems, in *IEEE Global Communications Conference (GLOBECOM) Dallas, USA*, vol. 1, pp. 225–229 (2004)

10. R. Cendrillon, W. Yu, M. Moonen, J. Verliden, T. Bostoen, Optimal multi-user spectrum management for digital subscriber lines. IEEE Trans. Commun. **54**(5), 922–933 (2006). https://doi.org/10.1109/TCOMM.2006.873096

11. S. Shenker, Making greed work in networks: a game theoretic analysis of switch service disciplines. IEEE/ACM Trans. Netw. **3**(6), 819–831 (1995)

12. Y.A. Korilis, A. Lazar, A. Orda, Achieving network optima using Stackelberg routing strategies. IEEE/ACM Trans. Netw. **1**(5), 161–172 (1997)

13. A. Orda, R. Rom, N. Shimkin, Competitive routing in multiuser communication networks. IEEE/ACM Trans. Netw. **1**(5), 510–521 (1993)

14. D.M. Kreps, *Game Theory and Economic Modelling* (Clarendon Press, Oxford, 1990)

Chapter 14
Multi-Service Network Models

14.1 Introduction

In thinking about how price is determined, the first rationale is to set subsidy-free prices or sustainable prices. Setting up subsidy-free prices are important if an incumbent wants to protect itself from competitors. Cross-subsidization of a service A another service B will allow a potential competitor to produce only service B and sell it for a lower price. Telecommunication networks present a challenge because a large part of the total cost is common and it is not easy to rationally allocate the common cost among different services. Sustainable pricing requires absence of cross-subsidization.

A second rationale for setting prices is driven by the objective to match supply and demand in the marketplace. Prices for services are expected to match supply and demand. Too low price levels will likely result in unsatisfied demand. Likewise, if prices are too high, there will likely be an oversupply of services. The right level of pricing will precisely match supply and demand.

The two rationales discussed above for setting prices do not necessarily lead to the same prices. There is possibly no single recipe for setting prices that satisfies all possible requirements. Pricing may also depend on the context. For example, a monopoly supplier in a market sets price to maximize its profits. If this market is under regulation, a regulator may require prices to maximize the social welfare, which not only includes supplier's benefits, but also users' welfare. This means that the task of pricing requires a careful balance between customers' needs, their willingness to pay, the underlying factors driven by technology, as well as the regulatory environment.

In multi-service networks, performance of each class of service and users' satisfaction will be directly influenced by all users' service choices. In best-effort (single QoS) networks, it is also true that individual performance is affected by the characteristics of aggregate traffic. However, this interdependence among users is even more complicated in multi-service networks. In this chapter, traffic from each

P. Verma, F. Zhang, *The Economics of Telecommunication Services*, Textbooks in Telecommunication Engineering, https://doi.org/10.1007/978-3-030-33865-7_14

class of service has different sensitivity to delay (disutility to average delay). The network resource allocation among all classes of users is modeled as a cooperative game to minimize the total disutility.

In addition, since each user's satisfaction (utility) is influenced not only by the network provider's pricing policy but also by all other users' service choices, we model this interdependence among users' service choices through a non-cooperative game theoretic framework. The service provider sets price for each class of service to maximize its revenue. Users purchase service from the service provider to maximize their utility functions independently under their budget constraints. The operating point is determined by the equilibrium where the price for each class of service is market-clearing price.

In this chapter, our basic assumptions regarding users' behaviors, as well as their utility functions and service provider's objective, are discussed in Sect. 14.2. Section 14.3 contains assumptions made about the network, while in Sect. 14.4 we discuss a general pricing model. We close with a brief summary of this chapter in Sect. 14.5.

14.2 Users and Network Provider Models

The pricing game we study in this chapter has two players: the network provider and finite set of network users. We develop a price for each class of service. Based on these prices, each user makes her service choices. Users do not cooperate with one another when deciding on an optimal strategy (hence characterizing a non-cooperative game), but rather each user acts individually striving to maximize her own payoff function (utility function).

14.2.1 Utility Functions for Users

Utility functions represent users' preferences, especially their sensitivity to QoS and level of demand for network services. In the context of this book, it is useful to think of utility as users' willingness to pay for a certain resource available to them.

Through users' actions, we can sometimes assess their willingness to pay for certain improvements in quality. A complete characterization of actual utility function is unlikely in practice. However, it is generally reasonable to assume that user i's utility function u_i possesses the following properties:

Assumption 14.1 (Allocated Bandwidth and QoS) *The user i's utility or u_i depends on the bandwidth allocated to user i and the received QoS.*

In networks, a user's utility is a function of allocated bandwidth and the QoS received. For example, a fixed bandwidth with a lower average delay will be preferred by the user over the same bandwidth with a higher average delay.

Assumption 14.2 (Monotonicity) u_i *is a monotonic function of its variables.*

The validity of this assumption is obvious. For example, the lower the average delay, the higher the perceived utility by the user. Monotonicity does not need to be strict; the perceived increase in utility as a function of bandwidth might stop at a certain level of bandwidth.

Assumption 14.3 (Concaveness) u_i *is concave function of its variables.*

This assumption is intuitive. The marginal utility almost always decreases with increasing levels of bandwidth and QoS [1].

Combinations of above assumptions are adopted in many pricing studies that employ the concept of utility functions, including [2–6].

The well-known utility function in data communication networks, proposed by Kelly [2], has the form $u_i = w_i \log(x_i)$, where w_i is user's willingness to pay and x_i is the allocated bandwidth. In this chapter, we define utility as follows:

$$u = \frac{\beta}{T_{now}} \log\left(\frac{x}{\tilde{x}}\right) \tag{14.1}$$

where $\beta > 0$ is a weighting factor. It characterizes the flow's relative sensitivity to the QoS parameter based on the fact that applications exhibit varying degree of sensitivity to QoS parameters (here we use delay as the QoS parameter for the network). For example, real-time voice and video are very sensitive to delay; packets that do not arrive within some delay bound cannot be used for playback and are in effect considered lost (although there are ways to make these applications less sensitive to such losses using coding or extra buffer). On the other hand, traditional data applications such as email service or file transfer are typically not very sensitive to delay. Thus, we use β to denote applications' QoS sensitivity characteristic.

In this book, we consider both elastic and inelastic users as defined by Shenker [7]. Traditionally, real-time voice and video applications that employ constant bit rate coding with no tolerance to eventual packet losses require a fixed amount of bandwidth for adequate QoS. There exist a variety of ways in which real-time applications can tolerate variations bandwidth using coding technologies and interpolation of the received data; however, some minimum bandwidth is nevertheless required. Although traditional data applications are elastic in nature and tend to be tolerant of variations in delay and can take advantage of even minimal amount of bandwidth, to guarantee users' network experience, we still assume a minimum bandwidth requirement. Here in Eq. (14.1) we use $\tilde{x}$ to denote this minimum bandwidth requirement. Notice that we use the *present* average delay T_{now} to represent the QoS parameter of the network. The reason for this choice is that users can easily track the present network situation such as Round-Trip Time, packet loss rate, etc. We can easily calculate T_{now} from these parameters.

14.2.2 Utility Functions for the Network Provider

Our model treats the network provider as a monopolist, a common assumption employed by most recent pricing studies discussed in Sect. 8.5.3.

In order to deter future competitors, it is better for a network provider to set up subsidy-free price for each class of service. Due to statistical multiplexing in networks as discussed in Sect. 8.3, each class of traffic will incur a waiting cost c_i based on delay. The utility function for the network is then defined as the sum of costs incurred by all classes and the network provider will try to minimize this total waiting cost:

$$u_s = \sum_{i=1}^{n} c_i \tag{14.2}$$

Alternatively, the network provider may use revenue as its utility function [3, 8]. When we use a vector p to denote the price of each class of service and vector s denote the supply of each class of service, the utility function for the service provider in this situation is defined as

$$u_s = p \times s \tag{14.3}$$

In this section, we have discussed the general forms of utility functions of both users and service provides.

14.3 Network Model

Here, we consider two different network models. The first model consists of a single source-destination pair common to all users. The study of a single queue is applicable to local and metropolitan area networks (e.g., Expedited Forwarding Per-Hop behavior traffic [9]), which are sometimes modeled as a single server with a queue that is distributed among all stations. This common source-destination model is also employed in [3, 10, 11].

In the second model, each class has a promised QoS and maintains a separate queue. The network resource is allocated among these classes using scheduling schemes like Weighted Round Robin (WRR), Class-based Queuing (CBQ) [12], and dynamic WRR [13]. This class-based network structure has been used in [5, 13–15].

Routing is assumed to be fixed and independent of the pricing policy. These simplifying assumptions free us from routing concerns. In a real commercial network, we find that routing will not play a fundamental role in the pricing issue since it is unlikely that the service provider will ask users to pay different prices depending on the actual route, given the same QoS level.

A non-preemptive priority-based multi-class system (single queue) will be used in Chap. 15 and a class-based network architecture (separate queue for each class) will be examined in Chaps. 16 and 17.

14.4 Pricing Model

The amount charged for network services may be a function of the combination of several factors, most notably as follows.

- **Access cost**: A charge for accessing the network services.
- **Service type**: In networks with multiple service categories, each class of service will be priced differently to reflect the QoS it provides.
- **Usage charge**: the usage charge is determined by the level of service provided to user and the actual amount of traffic consumed by user. Usage-based charge component can be used to discourage over-consumption and provide better network performance.
- **Time-of-day sensitive charge**: The time of consumption of network resources is relevant when implementing time-of-day pricing. The main objective of time-of-day pricing is to produce smoothing of traffic by encouraging users to shift their demand to times when the network is more lightly loaded.

In Table 14.1, we list the components of pricing of several studies on network pricing, as discussed in Chap. 8. We combine the factors discussed above into a general pricing policy model and the charge P_{ij} for class of service j to user i according to the following expression:

$$P_{ij} = c_j + p_j(t) * x_{ij} \tag{14.4}$$

In Eq. (14.4), c_j is a fixed access charge assigned to service j and x_{ij} is the amount of service j consumed by user i. The unit price of service j, p_j, is dependent on time t. This makes the model general enough to encompass time-of-day pricing.

In this book, we attempt to find the desired price vector $p = (p_1, p_2, \ldots, p_j, \ldots, p_m)$ for the service provider that supports m classes of services under pricing contexts described in the beginning of this chapter.

Table 14.1 Pricing mechanisms employed in various studies

	[16]	[10]	[6]	[5]	[14]
Access cost		×		×	
Service type	×	×	×	×	
Usage charge		×	×	×	×
Time-of-day sensitive charge	×				

14.5   Summary

This chapter has modeled a user's preferences through a utility function expressed in terms of the QoS parameter such as delay and the allocated bandwidth. Utility functions are assumed to be monotonic and concave. Each user will independently choose a strategy with the objective of maximizing his or her own utility function under his or her budget constraint. The network service provider is assumed to be a monopolist with the aim of either minimizing the total waiting cost or maximizing its revenue. Two different network models and a general form of pricing policy have been considered.

Problems

14.1 Give examples of concave functions.

14.2 The amount charged by network providers may be a function of several factors. Name a few factors.

References

1. A. Watson, M.A. Sasse, Evaluating audio and video quality in low-cost multimedia conferencing systems. Interact. Comput. **8**(3), 255–275 (1996)
2. F. Kelly, A. Maulloo, D. Tan, Rate control in communication networks: shadow prices, proportional fairness and stability. J. Oper.Res. Soc. **49**, 237–252 (1998)
3. M.L. Honig, K. Steiglitz, Usage-based pricing of packet data generated by a heterogeneous user population, in *Proceedings of the IEEE INFOCOM*, vol. 2 (IEEE, Boston, 1995), pp. 867–874
4. H. Ji, J.Y. Hui, E. Karasan, GoS-based pricing and resource allocation for multimedia broadband networks, in *Proceedings of the IEEE INFOCOM, San Francisco, CA* (1996), pp. 1020–1027
5. X. Wang, H. Schulzrinne, Pricing network resources for adaptive applications. IEEE/ACM Trans. Netw. **14**(3), 506–519 (2006)
6. H. Yaiche, R. Mazumdar, C. Rosenberg, A game theoretic framework for bandwidth. Allocation and pricing in broadband networks. IEEE/ACM Trans. Network. (TON) **8**(5) (2000)
7. S.J. Shenker, Fundamental design issues for the future internet. IEEE J. Sel. Areas Commun. **13**(7), 1176–1188 (1995)
8. F.P. Kelly, Charging and rate control for elastic traffic. Eur. Trans. Commun. **8**, 33–37 (1997)
9. B. Davie, A. Charny, *RFC 3246, An Expedited Forwarding PHB (Per-Hop Behavior)* (2002) DOI: 10.17487/RFC3246
10. L.A. Dasilva, D.W. Petr, N. Akar, Equilibrium pricing in multi-service priority-based networks, in *Proceedings of the IEEE/Global Telecommunications Conference (GLOBECOM)*, vol. 3 (1997), pp. 1373–1377
11. Y.A. Korilis, A. Lazar, A. Orda, Achieving network optima using Stackelberg routing strategies. IEEE/ACM Trans. Netw. **1**(5), 161–172 (1997)

12. G. Mamais, M. Markaki, G. Politis, I.S. Venieris, Efficient buffer management and scheduling in a combined IntServ and DiffServ architecture: a performance study, in *Proceedings of the ICATM* (1999), pp. 236–242
13. S. Yi, X. Deng, G. Kesidis, C.R. Das, Providing fairness in DiffServ architecture, in *Global Telecommunications Conference*, vol. 2 (2002), pp. 1435–1439
14. F. Zhang, P. Verma, A constant revenue model for packet switched network, in *IEEE GIIS 09, Hammamet, Tunisia* (2009)
15. F. Zhang, P.K. Verma, A two-step quality of service provisioning in multi-class networks, in *18th International Conference on Telecommunications (ICT 2011), Ayia Napa, Cyprus* (2011)
16. R. Cocchi, S. Shenker, D. Estrin, L. Zhang, Pricing in computer networks: motivation, formulation, and example. IEEE/ACM Trans. Netw. **1**(6), 614–627 (1993)

Chapter 15
Subsidy-Free Prices in Priority-Based Networks

15.1 Introduction

This chapter considers price differences among different classes as a cooperative queuing problem. Each class of service has a different waiting cost per unit of time (waiting cost factor) [1]. A cooperating queue is organized to minimize the total waiting cost of all classes while monetary compensations are set up for those classes which have to wait longer time. It is reasonable to assume this queuing problem as a transfer utility game and solve it by applying the Shapley Value. We consider the total cost of a coalition as the total waiting cost its members (classes) would incur if they had the power to be served first. The waiting cost associated with each class corresponds to the Shapley Value of the queuing game. In this chapter, we also find the solutions associated with the Shapley Value which satisfy many fairness properties like any classes which are served before another class are responsible to compensate the latter for their waiting cost; the sum of all compensation transfers is equal to zero.

Using Shapley Value to deal with the cost sharing queuing problem has been considered in [2–6]. In [2], the author has studied monetary transfer between agents based on a model where each agent has a different unit waiting cost. For such a model, this paper has characterized the Shapley Value rule using classical fairness axioms. Reference [3] has interpreted the worth of a coalition of agents in a different manner for the same model as in [2], and derived a different rule. It has also characterized this different rule using similar fairness axioms. In [4, 5], the queuing problem is studied from a strategic point of view under the assumption that all agents have identical unit waiting cost. The study in [6] is also based on the same model described in [2], and has considered cost sharing when both unit waiting cost and processing time of agents are present.

In this chapter, we extend the model described in [2] to multi-class communication networks and add the stochastic property of the communication network to it.

© Springer Nature Switzerland AG 2020
P. Verma, F. Zhang, *The Economics of Telecommunication Services*, Textbooks in Telecommunication Engineering, https://doi.org/10.1007/978-3-030-33865-7_15

We also propose a fair and efficient way to get the price difference among different classes of service.

The rest of this chapter is organized as follows. The model adopted for priority service is discussed in Sect. 15.2. The queuing games and the Shapley Value are studied in Sect. 15.3. In Sect. 15.4, we define the fairness axioms and check our results with these axioms. An illustrative example is given in Sects. 15.5 and 15.6 captures our conclusions.

15.2 The Model

Figure 15.1 shows the model provided in this chapter. A single communication server with a defined capacity c represents the packet switched network. A non-preemptive priority scheme is assumed. There are n different classes of service in the network. The packet length for all classes is considered to have the same statistics, and an exponential distribution with the average packet length equal to $\frac{1}{\mu}$ is assumed.

The set of classes are denoted as $N = \{1, \ldots, n\}$. $\sigma : N \to N$ is priority ordering of n classes of service and σ_i denotes the priority of class i. Each class i is distributed with Poisson arrivals and identified by two parameters: (λ_i, θ_i). λ_i is the average arrival rate and θ_i is the waiting cost factor for class i. A queuing problem is defined by a list $q = (N, \lambda, \theta) \in Q$, where Q is the set of all possible lists. Given an ordering σ, the waiting cost incurred by class i is given by

$$v_i(\sigma) = \lambda_i w_i \theta_i \tag{15.1}$$

where w_i is the average waiting time of class i packets given the ordering σ.

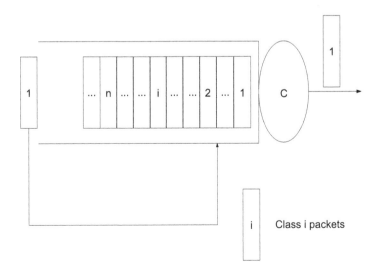

Fig. 15.1 FIFO non-preemptive priority schedule

The total waiting cost incurred by packets from all classes given an ordering σ can be written as

$$v(N, \sigma) = \sum_{i=1}^{n} v_i(\sigma) = \sum_{i=1}^{n} \lambda_i w_i \theta_i \tag{15.2}$$

Kleinrock [7] presents closed-form results for the waiting time in a non-preemptive priority discipline for a single M/G/1 queue, which we can apply here. We use $\bar{x}_i, \bar{x}_i^2$ to denote the first two moments of service time for class i packets. We have already assumed that all packets have identical statistical distribution with the average packet length $\frac{1}{\mu}$, therefore, $\bar{x}_i = \frac{1}{\mu c}, \bar{x}_i^2 = \frac{2}{(\mu c)^2}, \forall i = 1, \ldots, n$.

First, we define

$$w_0 = \sum_{i=1}^{N} \frac{\lambda_i \bar{x}_i^2}{2} = \frac{\sum_{i=1}^{n} \lambda_i}{(\mu c)^2} \tag{15.3}$$

We also define $\lambda^i = \bar{x}_i \sum_{j=1}^{i} \lambda_i = \frac{\sum_{j=1}^{i} \lambda_i}{\mu c}$, and $\lambda^0 = 1$. The waiting time for packets in class i is [7]

$$w_i = \frac{w_0}{(1 - \lambda^{i-1})(1 - \lambda^i)} \tag{15.4}$$

Here we define an allocation for a queuing problem $q = (N, \lambda, \theta) \in Q$ as $\psi(\sigma, t)$, where σ is an ordering, and t_i is the transfer related to class i packets. Given an ordering σ and a transfer t_i, the waiting cost share for class i packets is defined as,

$$u_i = v_i(\sigma) + t_i = \lambda_i \theta_i w_i + t_i \tag{15.5}$$

We define an efficient allocation as follows: An allocation $\psi(\sigma, t)$ is efficient for queuing problem $q = (N, \lambda, \theta) \in Q$ whenever it minimizes, i.e., the total cost of waiting minimizes $v(N, \sigma)$, and the algebraic sum of transfer is equal to 0, $\sum_{i=1}^{n} t_i = 0$.

An efficient ordering σ^* for the queuing problem $q = (N, \lambda, \theta) \in Q$ is the one which minimizes the total waiting cost $v(N, \sigma^*)$ incurred by packets from all classes. It means that $v(N, \sigma^*) \le v(N, \sigma), \forall \sigma$. For notational simplicity, we will write the total waiting cost in the efficient ordering of classes from N as $v(N)$, whenever it is not confusing. In some cases, we will deal with only a subset classes $S \subseteq N$. The ordering σ will then be defined on the classes in S only, and we will write the total waiting cost from an efficient ordering of classes in S as $v(S)$.

The following lemma shows that when classes are ordered in decreasing θ, it is an efficient ordering.

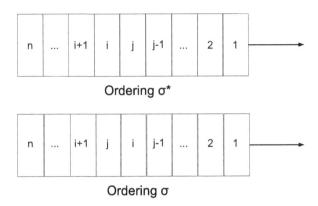

Fig. 15.2 Ordering of n classes

Lemma 15.1 *For any* $S \subseteq N$, *let* σ^* *be an efficient ordering of classes in S. For every* $i \neq j$, $i, j \in S$, *if* $\sigma_i^* > \sigma_j^*$, *then* $\theta_i < \theta_j$.

Proof If we assume the contrary, it will imply that two consecutive classes, i, $j \in S(\sigma_i = \sigma_j + 1)$ can be found such that $\theta_i > \theta_j$. We can then define a new ordering σ as shown in Fig. 15.2 by interchanging i and j in σ^*.

As shown in Eq. (15.4), the average waiting time for classes in $S \setminus \{i, j\}$ is not changed from ordering σ^* to σ. And based on Eq. (15.1), the costs to classes in $S \setminus \{i, j\}$ also remain unchanged. Therefore, the difference between total costs in σ^* and σ is given by

$$v\left(S, \sigma^*\right) - v(S, \sigma) = \lambda_i w_i^{\sigma^*} \theta_i + \lambda_j w_j^{\sigma^*} \theta_j - \left(\lambda_j w_j^{\sigma} \theta_j + \lambda_i w_i^{\sigma} \theta_i\right) \qquad (15.6)$$

As shown in Fig. 15.2, we can get the average waiting time for class i, j packets in ordering σ^* and σ as follows:

$$w_i^{\sigma^*} = \frac{w_0}{\left(1 - \lambda^{j-1} - \frac{\lambda_j}{\mu c}\right)\left(1 - \lambda^{j-1} - \frac{\lambda_j}{\mu c} - \frac{\lambda_i}{\mu c}\right)}$$

$$w_j^{\sigma^*} = \frac{w_0}{\left(1 - \lambda^{j-1}\right)\left(1 - \lambda^{j-1} - \frac{\lambda_j}{\mu c}\right)}$$

$$w_i^{\sigma} = \frac{w_0}{\left(1 - \lambda^{j-1}\right)\left(1 - \lambda^{j-1} - \frac{\lambda_i}{\mu c}\right)}$$

$$w_j^{\sigma} = \frac{w_0}{\left(1 - \lambda^{j-1} - \frac{\lambda_i}{\mu c}\right)\left(1 - \lambda^{j-1} - \frac{\lambda_i}{\mu c} - \frac{\lambda_j}{\mu c}\right)}$$

Now, we take the above equations into Eq. (15.6), we get

$$v(S, \sigma^*) - v(S, \sigma)$$

$$= \frac{\lambda_i \lambda_j (\theta_i - \theta_j) \left(1 - \lambda^{j-1} - \frac{\lambda_i}{\mu c} + 1 - \lambda^{j-1} - \frac{\lambda_j}{\mu c}\right)}{(1 - \lambda^{j-1}) \left(1 - \lambda^{j-1} - \frac{\lambda_i}{\mu c}\right) \left(1 - \lambda^{j-1} - \frac{\lambda_j}{\mu c}\right) \left(1 - \lambda^{j-1} - \frac{\lambda_i}{\mu c} - \frac{\lambda_j}{\mu c}\right)}$$

We have already assumed that σ^* is an efficient ordering, and we get $v(S, \sigma^*) - v(S, \sigma) \le 0$, this gives us $\theta_i \le \theta_j$, which is a contradiction.

This proves Lemma 15.1.

Notice that the efficient queuing problem is independent of the transfer and is unique when all classes have different unit waiting cost. And we can rewrite that an allocation $\psi(\sigma, t_i)$ is efficient for the queuing problem $q = (N, \lambda, \theta) \in Q$ whenever σ is an efficient ordering and $\sum_{i=1}^{n} t_i = 0$.

From Eq. (15.5), we can calculate the actual waiting cost for each class $v_i(\sigma)$ based on efficient ordering as described in Lemma 15.1. In the next section, we will consider the waiting cost share problem as a cooperative game and set up the waiting cost share for each class u_i using the Shapley Value. For each class, if we know the actual waiting cost $v_i(\sigma)$ and waiting cost share u_i, based on Eq. (15.5), we can find the transfer t_i for each class.

The inequality $t_i < 0$ shows that class i will receive compensation and $t_i > 0$ shows it will compensate other classes. After getting the transfer t_i between different classes, we are able to define the price difference Δp_{ij} between class i and j as follows:

$$\Delta p_{ij} = \frac{t_i}{\lambda_i} - \frac{t_j}{\lambda_j} \tag{15.7}$$

15.3   Waiting Cost Sharing Using the Shapley Value

As discussed in Sect. 15.2, we solve the waiting cost share queuing problem by treating it as a cooperative game. In this section, we first define the coalitional cost of this game and then analyze the solution based on Shapley Value of the corresponding game.

Given a queue $q \in Q$, the waiting cost of a coalition of $S \subseteq N$ classes in the queue is defined as the cost incurred by the classes in S if they have the power to be served first in the queue and use an efficient ordering in S.

And, the value of a coalition $S \subseteq N$ is as follows:

$$v(S) = v(S, \sigma) = \sum_{i \in S} \lambda_i w_i \theta_i \tag{15.8}$$

where $\sigma = \sigma(S)$ is an efficient ordering considering classes in S only. In [2], the author also studied another equivalent way to define the worth of a coalition using the dual function of the cost function. Other ways to define the worth of a coalition are addressed in [3] which assumed that a coalition of classes is served after the classes not in the coalition.

The marginal contribution of class $i \in N$ to a coalition S in $v(S)$, $i \notin S$ is a sum of the costs associated with each member of S. Indeed, those classes having a higher unit waiting cost θ than class i impose a waiting cost on it, and those having a lower unit waiting cost θ than class i have to wait additional units of time. This implies that the marginal contribution is equal to the cost of waiting of class i itself, and the cost its existence imposes on those classes that follow it in the new queue. Stated mathematically, for $q = (N, \lambda, \theta) \in Q$, $S \subset N$, $i \in N \setminus S$, the marginal contribution of class i is

$$v(S \cup \{i\}) - v(S) = v(S \cup \{i\}, \sigma') - v(s, \sigma) = \sum_{i \in S \cup \{i\}} \lambda_i w_i^{\sigma'} \theta_i - \sum_{i \in S} \lambda_i w_i^{\sigma} \theta_i$$

(15.9)

where $\sigma' = \sigma(S \cup \{i\})$, $\sigma = \sigma(S)$ are efficient orderings considering classes in $S \cup \{i\}$ and S, respectively.

The Shapley Value (waiting cost share) of class i is defined as a weighted sum of the class's marginal contribution to coalitions. Based on the definition of the Shapley Value as described in Sect. 13.1.2, for all $q = (N, \lambda, \theta) \in Q$, $i \in N$, the payoff assigned to class i is given by

$$u_i = SV_i = \sum_{S \subseteq N \setminus \{i\}} \frac{|S|!(|N| - |S| - 1)!}{|N|!} [v(S \cup \{i\}) - v(S)]$$

(15.10)

The Shapley Value allocation rule says that classes are ordered using an efficient ordering and transfers (compensations) are assigned to classes such that the cost share of each class is equal to its Shapley Value. Based on the efficiency property of Shapley Value described in Sect. 13.1.2, the total gain is distributed among N players, that is, $\sum_{i \in N} SV_i = $ minimum $v(N, \lambda, \theta)$.

Another way to write the Shapley Value formula is as follows [8]:

$$SV_i = \sum_{S \subseteq N : i \in S} \frac{\Delta(S)}{|S|}$$

(15.11)

where $\Delta(S) = v(S)$ if $|S| = 1$ and $\Delta(S) = v(S) - \sum_{T \subset S} \Delta(T)$. This gives $\Delta(\{i\}) = v(\{i\}) = \lambda_i w_i^{\{i\}} \theta_i$, $\forall i \in N$, $w_i^{\{i\}}$ is the waiting time for class i when class i packets have the power to be served first in the queue. For any $i, j \in N$, that is, $|S| = 2$ we have

$$\Delta(S) = \Delta\{i, j\} = v(\{i, j\}) - v(\{i\}) - v(\{j\})$$
$$= \lambda_i w_i^{\{i,j\}} \theta_j + \lambda_j w_j^{\{i,j\}} \theta_j - \lambda_i w_i^{\{i\}} \theta_i - \lambda_j w_j^{\{j\}} \theta_j$$

We assume $\theta_i \geq \theta_j$ and consider Eq. (15.4), we have,

$$\Delta(S) = \Delta\{i, j\} = v(\{i, j\}) - v(\{i\}) - v(\{j\}) = \lambda_j \theta_j \left(w_j^{\{i,j\}} - w_j^{\{j\}} \right)$$

If $|S| = 3$, say $S = \{i, j, k\}$, then,

$$\begin{aligned}
\Delta(S) &= \Delta\{i, j, k\} \\
&= v(\{i, j, k\}) - \Delta(\{i, j\}) - \Delta(\{j, k\}) - \Delta(\{i, k\}) - \Delta(\{i\}) - \Delta(\{j\}) - \Delta(\{k\})
\end{aligned}$$

If we further assume $\theta_i \geq \theta_j \geq \theta_k$, we have,

$$\Delta(S) = \Delta\{i, j, k\} = \lambda_k \theta_k \left(w_k^{\{i,j,k\}} - w_k^{\{i,k\}} - w_k^{\{j,k\}} + w_k^{\{k\}} \right)$$

It is easy to use induction to show that when $S = \{1, 2, \ldots, n\}$ with $\theta_1 \geq \theta_2 \geq \ldots \geq \theta_n, \sigma_1 = 1, \sigma_2 = 2, \ldots, \sigma_n = n$ (we will use i to denote the position of each class i in the queue instead of σ_i for simplicity) $\Delta(S)$ is

$$\Delta(S) = \Delta\{1, 2, \ldots, n\} = \lambda_n \theta_n w_n^{\Delta(S)} \tag{15.12}$$

where $w_n^{\Delta(S)} = \sum_{T \subseteq S, n \in T} (-1)^{|T|} w_n^T$, when $|S|$ is even, and $w_n^{\Delta(S)} = \sum_{T \subseteq S, n \in T} (-1)^{|T|+1} w_n^T$, when $|S|$ is odd. We can see that the $\Delta(S)$ only depends on the lowest priority class, that is, $\lambda_n \theta_n w_n^{\Delta(S)}$.

Now, we are ready to consider $SV_i = \sum_{S \subseteq N : i \in S} \frac{\Delta(S)}{|S|}$ in more detail. Given a length of a set $|S|$, $S \subseteq N$, $i \in S$, there are $\binom{i-1}{|S|-1}$ situations where class i is the lowest priority class in set S. Denote $\mathcal{A}_i$ as the set of S satisfies the aforementioned situations. Similarly, for class j which has the position $j > i$, there are $\binom{j-2}{|S|-2}$ situations where class j is the lowest priority class in S. Denote $\mathcal{B}_j$ as the set of S satisfies the situations. Therefore, we can rewrite Eq. (15.11) as

$$SV_i = \sum_{S \subseteq N, i \in S} \frac{\sum_{S' \in \mathcal{A}_i} \lambda_i \theta_i w_i^{\Delta(S')} + \sum_{j=i+1}^{n} \sum_{S' \in \mathcal{B}_j} \lambda_j \theta_j w_j^{\Delta(S')}}{|S|} \tag{15.13}$$

Using Eq. (15.13), we can also show the efficiency property of the Shapley Value, i.e., $\sum_{i=1}^{n} SV_i = \sum_{i=1}^{n} \lambda_i \theta_i w_i^{\{1,\ldots,n\}} = \sum_{i=1}^{n} v_i(\sigma) = v(N, \sigma)$.

After we get SV_i, the transfer t_i for each class can be calculated as follows:

$$t_i = SV_i - v_i(\sigma) = SV_i - \lambda_i \theta_i w_i^{\{1,\ldots,n\}} \tag{15.14}$$

Lemma 15.2 *Using the Shapley Value as the waiting cost share for each class, the allocation $\psi(\sigma, t)$ is efficient.*

Proof We already stated the efficient allocation definition in Sect. 15.2: if an allocation $\psi(\sigma, t)$ for queuing problem $q = (N, \lambda, \theta) \in Q$ minimizes the total waiting cost $v(N, \sigma)$ and no transfer is lost ($\sum_{i=1}^{n} t_i = 0$), then this allocation is efficient.

First, from the efficiency property of the Shapley Value, we have $\sum_{i=1}^{n} SV_i =$ minimize $v(N, \sigma)$.

From Eq. (15.14), we know that: $\sum_{i=1}^{n} t_i = \sum_{i=1}^{n} SV_i - \sum_{i=1}^{n} v_i(\sigma)$.

The σ in Eq. (15.14) is an efficient ordering and from Lemma 15.1, $\sum_{i=1}^{n} v_i(\sigma)$ is the minimum system cost, minimum $v(N, \sigma)$.

Therefore, we have, $\sum_{i=1}^{n} t_i =$ minimum $v(N, \sigma) -$ minimum $v(N, \sigma) = 0$.

This proves Lemma 15.2.

Taking Eq. (15.14) into Eq. (15.7), we get the complete formation of price difference between class i and class j as follows:

$$\Delta p_{ij} = \frac{SV_i - \lambda_i \theta_i w_i^{\{1,\dots,n\}}}{\lambda_i} - \frac{SV_j - \lambda_j \theta_j w_j^{\{1,\dots,n\}}}{\lambda_j} \tag{15.15}$$

We have thus developed the subsidy-free price difference between classes based on the inter-class compensations. Since a network generally maintains a limited number of classes of service, the calculation of waiting cost share SV_i and actual waiting cost c_i for each class does not suffer from the scalability problem.

15.4   Axiomatic Characterization of the Shapley Value

As shown in Sect. 15.3, the price difference between classes directly depends on the waiting cost share rule in the network, specifically, using the Shapley Value as the waiting cost share for each class. In this section, we define several axioms on fairness and characterize the Shapley Value using them.

Definition 15.1 The waiting cost sharing rule satisfies the efficiency rule if and only if for all $q = (N, \lambda, \theta)$, $\psi(\sigma, t)$ is efficient.

As shown in Lemma 15.2, when we use Shapley Value as the waiting cost sharing for each class, $\psi(\sigma, t)$ is efficient.

The next definition is as in literature. For example, two similar classes should be compensated such that their cost shares are equal (equal treatment of equals).

Definition 15.2 The waiting cost sharing rule satisfies equal treatment of equals if the following condition is satisfied: For all $q = (N, \lambda, \theta) \in Q$, $\psi(\sigma, t)$, $i, j \in N$, then $\lambda_i = \lambda_j$, $\theta_i = \theta_j \Rightarrow u_i = u_j$.

Using Shapley Value as the waiting cost share for each class obviously satisfies equal treatment of equals axiom from Eq. (15.10).

Assume that the impatience of class i increases. Under this assumption, the total cost of waiting may increase. The following axiom, called independence of preceding classes impatience (IPAI), proposes that classes being served after class i be not affected by the increase. Since those classes are served after class i, they do not impose any cost of waiting to class i. It also implies that only the classes preceding class i impose a cost of waiting to class i. These classes must bear the consequences of an increase in that waiting cost.

Definition 15.3 The waiting cost sharing rule satisfies independence of preceding classes impatience (IPCI) if and only if for all $q = (N, \lambda, \theta)$, $q' = (N, \lambda, \theta') \in Q$, $\psi(\sigma, t)$, $\psi(\sigma', t')$, and for all $i \in N$, $\lambda_i = \lambda'_i$, $i \in N \backslash k : \theta_i = \theta'_i$ and $\theta_k < \theta'_k$, then for all $j \in N$ such that $\sigma_j > \sigma_k : u_j = u'_j$.

The proof using Shapley Value as the waiting cost share for each class which satisfies IPCI is given here.

Given that $q = (N, \lambda, \theta)$, $q' = (N, \lambda, \theta') \in Q$, $\psi(\sigma, t)$, $\psi(\sigma', t')$, $k \in N$, and for all $i \in N \backslash k : \theta_i = \theta'_i$ and $\theta_k < \theta'_k$, we get

- $\sigma_k \geq \sigma'_k$ and for any $j \in N \backslash k, \sigma_j > \sigma_k, \theta_j = \theta'_j$, and have the same ordering.

From Eq. (15.13), given a length of a set $|S|$, $S \subseteq N, j \in S$, there are $\binom{j-1}{|S|-1}$ situations where class j is the lowest priority class in set S. Denote $\mathcal{A}_j$ as the set of S satisfies the aforementioned situations. For class i which has the position $i > j$, there are $\binom{i-2}{|S|-2}$ situations where class i is the lowest priority class in S. Denote $\mathcal{B}_j$ as the set of S satisfies the situations. Therefore, we have

$$SV_j = \sum_{S \subseteq N, j \in S} \frac{\sum_{S' \in \mathcal{A}_j} \lambda_j \theta_j w_j^{\Delta(S')} + \sum_{i=j+1}^{n} \sum_{S' \in \mathcal{B}_i} \lambda_i \theta_i w_i^{\Delta(S')}}{|S|}$$

Similarly, given a length of a set $|S|$, $S \subseteq N$, $j \in S$, there are $\binom{j'-1}{|S|-1}$ situations where class j' is the lowest priority class in set S. Denote $\mathcal{A}'_j$ as the set of S satisfies the aforementioned situations. For class i' which has the position $i' > j'$, there are $\binom{i'-2}{|S|-2}$ situations where class i' is the lowest priority class in S. Denote $\mathcal{B}_j$ as the set of S satisfies the situations. Therefore, we have

$$SV'_j = \sum_{S \subseteq N, j' \in S} \frac{\sum_{S' \in \mathcal{A}'_j} \lambda'_j \theta'_j w_j^{'\Delta(S')} + \sum_{i'=j'+1}^{n} \sum_{S' \in \mathcal{B}'_i} \lambda'_i \theta'_i w_i^{'\Delta(S')}}{|S|}$$

We already have for all $i \in N$, $\lambda_i = \lambda'_i$. For any $j \in N$, when $\sigma_j > \sigma_k$, we have $\sigma_j = \sigma'_j$. From (15.4), we have $w_j = w'_j$. In addition, when we look at Eq. (15.12) and we can also find that $w_j^{\Delta(S)} = w_j^{'\Delta(S)}$, $S \subseteq N$. Thus, we conclude that $SV_j = SV'_j$ when $\sigma_j > \sigma_k$ and using Shapley Value as the waiting cost share for each class satisfies IPCI.

The last definition is consistent with the idea that if two classes are served before a third one (all three classes have identical traffic load λ), then the former are equal contributor to the waiting cost incurred by the latter. To capture this idea, we consider that the network no longer charges the last class (the unit waiting cost θ is 0). Then it is not necessary to change the queue. However, the transfer attributed to the last agent has to be redistributed among the remaining agents. This redistribution has to be equal among all because it will imply equal responsibility.

Definition 15.4 Waiting cost sharing rule satisfies equal responsibility (ER) if and only if for all $q = (N, \lambda, \theta)$, $\in Q$, $\psi(\sigma, t)$, if for all $i \in N$, $\lambda_i = \lambda$, $q' = (N, \lambda', \theta') \in Q$ such that for all $i \in N$: $\lambda'_i = \lambda$, for all $i \in N \backslash n$, $\theta'_i = \theta_i$, $\theta'_n = 0$, there exists $\psi(\sigma', t')$, such that for all $i \in M$:

- $\sigma'_i = \sigma_i$ and
- $t'_i = t_i + \frac{t_n}{n-1}$.

Using Shapley Value as the waiting cost share for each class satisfies ER. The proof follows.

Under Definition 15.4, for all $i \in N$, $\lambda_i = \lambda'_i = \lambda$, we have for $|S| = i$, $i = 1$, ..., n, $w^{\Delta(S)}$ is a constant and it is not related to the elements in S. For example, $|S_1| = |S_2| = 3$, and $S_1 = \{i, j, k\}$, $S_2 = \{a, b, c\}$ with $\theta_i \geq \theta_j \geq \theta_k$, $\theta_c \geq \theta_b \geq \theta_c$. From the $w^{\Delta(S)}$ described in Eq. (15.12), we have,

$$w_k^{\Delta(S_1)} = w_k^{\{i,j,k\}} - w_k^{\{i,k\}} - w_k^{\{j,k\}} + w_k^{\{k\}}$$

$$w_c^{\Delta(S_2)} = w_c^{\{a,b,c\}} - w_c^{\{a,c\}} - w_c^{\{b,c\}} + w_c^{\{c\}}$$

For all $i \in N$, $\lambda_i = \lambda$, from Eq. (15.4) we get,

$$w_k^{\Delta(S_1)} = w_c^{\Delta(S_2)} = \frac{w_0}{\left(1 - \frac{2\lambda}{\mu c}\right)\left(1 - \frac{3\lambda}{\mu c}\right)} - \frac{w_0}{\left(1 - \frac{\lambda}{\mu c}\right)\left(1 - \frac{2\lambda}{\mu c}\right)} - \frac{w_0}{\left(1 - \frac{\lambda}{\mu c}\right)\left(1 - \frac{2\lambda}{\mu c}\right)} + \frac{w_0}{1 - \frac{\lambda}{\mu c}}$$

For simplicity, in the following equation, we use $w^{|S|}$ to denote $w^{\Delta(S)}$ without differentiating among S as long they have same size.

From Eq. (15.13), we have,

$$SV_i = \sum_{|S|=1}^{n} \frac{\lambda w^{|S|}\left(\binom{i-1}{|S|-1}\theta_i + \sum_{j=i+1}^{n}\binom{j-2}{|S|-2}\theta_j\right)}{i}$$

In this case, the only difference between $q = (N, \lambda, \theta)$, $\psi(\sigma, t)$ and $q = (N, \lambda', \theta')$, $\psi(\sigma', t')$ is $\theta'_n = 0$. Therefore the efficient ordering σ' of q' is same as the efficient ordering σ of q ($\sigma'_i = \sigma_i$, for $i \in N$).

Since $\theta'_n = 0$, the Shapley Value for SV'_i is,

$$SV_i' = \sum_{|S|=1}^{n-1} \frac{\lambda w^{|S|} \left(\binom{i-1}{|S|-1} \theta_i + \sum_{j=i+1}^{n-1} \binom{j-2}{|S|-2} \theta_j \right)}{i}$$

From Eq. (15.5), we have,

$$t_i = SV_i - \lambda \theta_i w_i$$

$$t_i' = SV_i' - \lambda \theta_i' w_i'$$

Since for $i \in N$, $\lambda_i = \lambda_i' = \lambda$, we find $w_i = w_i' = \frac{w_0}{(1-\frac{(i-1)\lambda}{\mu c})(1-\frac{i\lambda}{\mu c})}$.

Therefore,

$$t_i - t_i' = SV_i - SV_i' = \sum_{|S|=1}^{n-1} \frac{\lambda w^{|S|} \binom{n-2}{|S|-2} \theta_n}{|S|}$$

From above equation we find that for $i = 1, \ldots, n-1$, $t_i - t_i'$ is equal to the identical value $\sum_{|S|=1}^{n-1} \frac{\lambda w^{|S|} \binom{n-2}{|S|-2} \theta_n}{|S|}$. In order to find out what this value stands for, we assume that

$$t_i' = t_i + x, i = 1, \ldots, n-1 \tag{15.16}$$

Since allocations $\psi(\sigma, t)$, $\psi(\sigma', t')$ are efficient, $\sum_{i=1}^{n} t_i = 0$, $\sum_{i=1}^{n-1} t_i' = 0$. Summation of both side of Eq. (15.16), we get,

$$\sum_{i=1}^{n-1} t_i' = \sum_{i=1}^{n-1} t_i + (n-1)x$$

Therefore, we get $x = \frac{t_n}{n-1}$, and $t_i' = t_i + \frac{t_n}{n-1}$. Note that we also find $\frac{t_n}{n-1} = -\sum_{|S|=1}^{n-1} \frac{\lambda w^{|S|} \binom{n-2}{|S|-2} \theta_n}{|S|}$.

Thus, we have proved that using Shapley Value as the waiting cost share for each class satisfies the ER axiom.

15.5 An Illustrative Example

This section presents a numerical example of the pricing scheme for the multi-class priority-based network proposed in this chapter.

To emphasize the methodology, we simply assume that the network has two different classes with the non-preemptive priority scheme. We use λ to denote the total traffic in the network and k the percentage of traffic choosing class one service which has θ_1 as the waiting cost factor. We assume $\theta_1 > \theta_2$. It means that class one packets have higher priority. Based on Lemma 15.1, the efficient ordering is $\sigma = \{1, 2\}$. Using Eq. (15.13), the waiting cost share for each class is as follows:

$$SV_1 = k\lambda\theta_1 w_1^{\{1\}} + \frac{(1-k)\lambda\theta_2 \left(w_2^{\{1,2\}} - w_2^{\{2\}}\right)}{2}$$

$$SV_2 = (1-k)\lambda\theta_2 w_2^{\{2\}} + \frac{(1-k)\lambda\theta_2 \left(w_2^{\{1,2\}} - w_2^{\{2\}}\right)}{2}$$

As described in Sect. 15.3, $w_i^S, i \in S$ is the average waiting cost of class i in an efficient ordering of S assuming that S has the power to be served first. Using Eq. (15.4), the average waiting time for each class in an efficient ordering can be calculated as

$$w_1^{\{1\}} = w_1^{\{1,2\}} = \frac{\lambda}{\mu c(\mu c - k\lambda)}$$

$$w_1^{\{2\}} = \frac{\lambda}{\mu c(\mu c - (1-k)\lambda)}$$

$$w_2^{\{1,2\}} = \frac{\lambda}{(\mu c - k\lambda)(\mu c - \lambda)}$$

As defined in Eq. (15.5), we can calculate the transfer for each class as follows:

$$t_1 = k\lambda\theta_1 w_1^{\{1,2\}} - SV_1 = -\frac{(1-k)\lambda\theta_2 \left(w_2^{\{1,2\}} - w_2^{\{2\}}\right)}{2}$$

$$t_2 = (1-k)\lambda\theta_2 w_2^{\{1,2\}} - SV_2 = \frac{(1-k)\lambda\theta_2 \left(w_2^{\{1,2\}} - w_2^{\{2\}}\right)}{2}$$

Using the definition in Eq. (15.7), the price difference between class one service and class two service is,

$$\Delta p_{12} = \frac{t_2}{(1-k)\lambda} - \frac{t_1}{k\lambda} = \frac{\theta_2 \left(w_2^{\{1,2\}} - w_2^{\{2\}}\right)}{2k}$$

For simplicity, we assume the network capacity c is equal to 100 and the average length of packets in the network $\frac{1}{\mu} = 1$. The unit waiting cost for each class as $\theta_1 = 1.5$ and $\theta_2 = 1$. Figure 15.3 describes $w_i, SV_i, t_i, \Delta p_{12}, (i = 1, 2)$ profiles against change λ total arrival rate in the network with fixed percentage of traffic choosing class one service k. In Fig. 15.3a, the average waiting time for both classes increases as the total average arrival rate increases while class one packets experiences much lower average delay. Figure 15.3b shows the waiting cost share (Shapley Value) for each class against the increasing total traffic arrival rate. Figure 15.3c describes the transfer of each class, specifically, the amount class one packets should compensate the class two packets. As shown in Sect. 15.2, total transfer is equal to 0, $t_1 + t_2 = 0$. Figure 15.3d shows that the price difference between the two classes increases as the total traffic in the network increases since class one packets need to pay higher price for the better service (lower average waiting time) when the network resource is short.

Figures shown in Fig. 15.3 profile the situations with fixed percentage of traffic choosing class one service, (k). And Fig. 15.4 shows all the $w_i, SV_i, t_i, \Delta p_{12}, (i = 1, 2)$ profiles against changing k with a given total arrival rate in the network (λ). Figure 15.4a shows that the average waiting time for both classes increases as the percentage k of packets choosing class one service. In Fig. 15.4b, the waiting cost share (Shapley Value) for each class increases as the percentage of traffic choosing that class increases. In other words, the waiting cost share for class one increases with fraction (k) of traffic choosing class one and the waiting cost share for class two increases with percentage of traffic choosing class two, $(1 - k)$. Figure 15.4c shows the amount class one packets should compensate class two packets as a function of k with the property $t_1 + t_2 = 0$. In Fig. 15.4d, the price difference between the two classes is minimized when each of the class has identical amount of traffic, i.e., $\lambda_1 = \lambda_2$. It makes sense in a way that when the percentage of traffic choosing class one (k) is small, the price difference between the two classes is higher since class one packets enjoy little economy of scale. And when the percentage of traffic choosing class one (k) is large, the class one packets need to pay more to compensate class two packets since class two packets experiencing much worse network quality (large average waiting time).

15.6 Summary

In this chapter, we have investigated the problem of pricing multi-class priority-based network services. Compensation is transferred from higher priority classes to lower priority classes which experience longer average waiting time. In the model proposed in this chapter, each class has Poisson arrivals and exponentially distributed packet length with identical mean, but a different waiting cost factor. We present an efficient ordering scheme by assigning each class its position in the queue, and then calculate compensation for each class based on its waiting cost share which is its Shapley Value in a cooperative game. We have also characterized

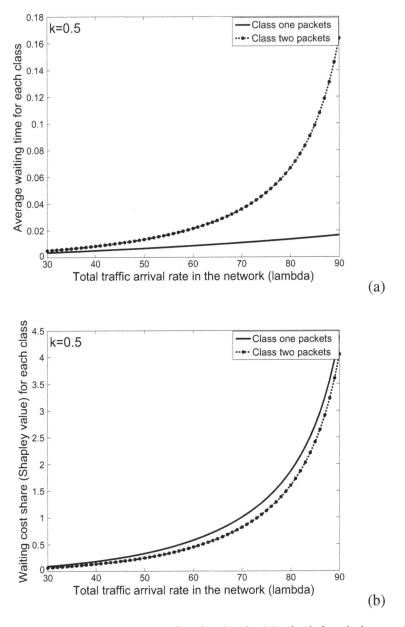

Fig. 15.3 Profiles w_i, SV_i, t_i, Δp_{12}, $(i = 1, 2)$ against changing λ (total arrival rate in the network) with fixed k (fraction of traffic choosing class one service). (**a**) Average waiting time w_i, $(i = 1, 2)$ for each class against λ. (**b**) Waiting cost share (Shapley Value) SV_i, $(i = 1, 2)$ for each class against λ. (**c**) Transfer t_i, $(i = 1, 2)$ for each class against λ. (**d**) Price difference between the two classes Δp_{12} against λ

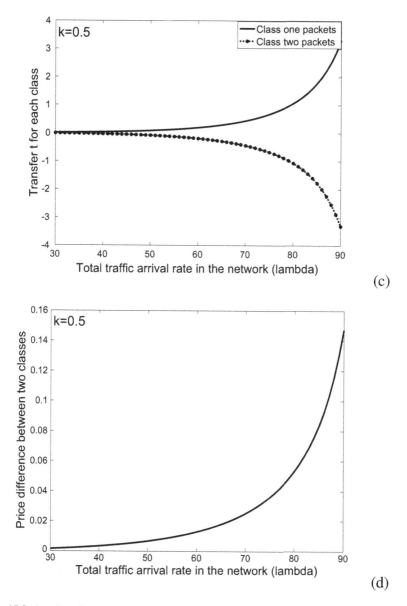

Fig. 15.3 (continued)

the Shapley Value using different intuitive fairness axioms. A numerical example illustrates how the analytical results presented in the chapter can be used in a practical situation.

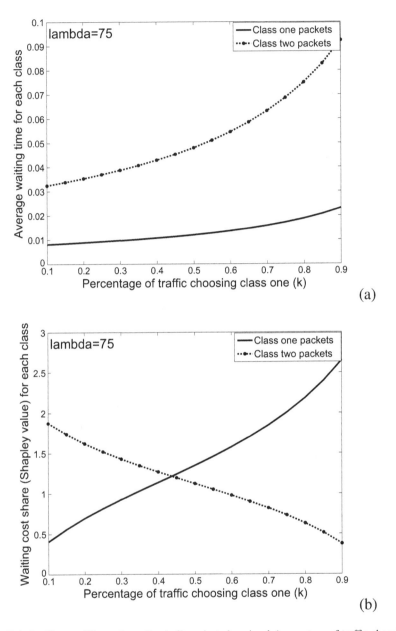

Fig. 15.4 Profiles w_i, SV_i, t_i, Δp_{12}, $(i = 1, 2)$ against changing k (percentage of traffic choosing class one service) with fixed λ (total arrival rate in the network). (**a**) Average waiting time w_i, $(i = 1, 2)$ for each class against k. (**b**) Waiting cost share (Shapley Value) SV_i, $(i = 1, 2)$ for each class against k. (**c**) Transfer t_i, $(i = 1, 2)$ for each class against k. (**d**) Price difference between the two classes Δp_{12} against k

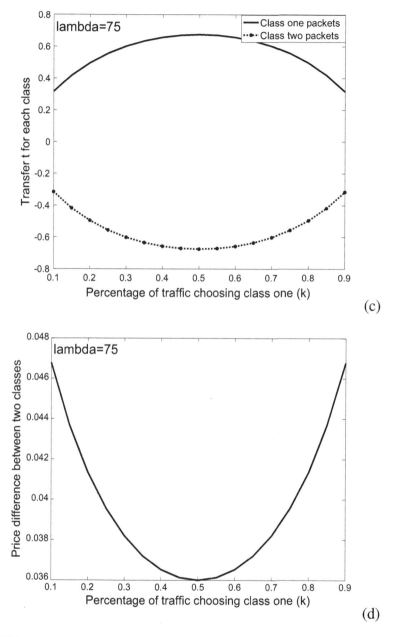

(c)

(d)

Fig. 15.4 (continued)

Problems

15.1 What is a non-preemptive priority queue?

15.2 In the model described in this chapter, if a network has 4 classes of service with the following waiting cost factors: $class_1 = 1$, $class_2 = 2$, $class_3 = 3$, $class_4 = 4$, what is an efficient ordering of the 4 classes of service?

15.3 In above 15.2 network setup, using Shapely value, find the waiting cost share for class 1 is sv_1, the actually waiting cost class 1 experienced is v_1, what is the transfer for class 1?

References

1. F. Zhang, P. Verma, *Pricing Class-based Networks Using the Shapley Value, the Journal of NETNOMICS* (2011). https://doi.org/10.1007/s11066-011-9058-5
2. F. Maniquet, A characterization of the Shapley value in queueing problems. J. Econ. Theory **109**, 90–103 (2003)
3. Y. Chun, *A Note on Maniquet's Characterization of the Shapley Value in Queueing Problems*, Working Paper (Rochester University, Rochester, 2004)
4. H. Moulin, *On Scheduling Fees to Prevent Merging, Splitting and Transferring of Jobs*, Working Paper (Rice University, Rice, 2004)
5. H. Moulin, *Split-proof Probabilistic Scheduling*, Working Paper, (Rice University, Rice, 2004)
6. D. Mishra, B. Rangarajan, Cost sharing in a job scheduling problem using the Shapley value, in *Proceedings of the 6th ACM conference on Electronic commerce, Vancouver, BC, Canada* (2005), pp. 232–239
7. L. Kleinrock, Queuing systems, in *Computer Applications*, vol. 2 (Wiley, New York, 1976)
8. J.C. Harsanyi, *Contributions to Theory of Fames IV, Chapter A Bargaining Model for Cooperative n-person Games* (Princeton University Press, Princeton, 1995)

Chapter 16
A Constant Revenue Model for Net Neutrality

16.1 Background

The lawsuit involving Comcast and BitTorrent in the USA brought widespread attention and the net neutrality debate polarized the major stakeholders, hardening their respective stands. On August 17, 2007, Comcast was reported to prevent BitTorrent users from seeding files [1]. Later, Comcast's limiting of Bit Torrent applications was further confirmed in a study conducted by the Electronic Frontier Foundation [2]. At the same time, Comcast argued that it considered choking BitTorrent traffic as a way to let the network traffic remain available for everyone. In January 2008, FCC Chairman Kevin Martin stated that the FCC was going to investigate complaints that Comcast "actively interferes with Internet traffic as its subscribers try to share files online" [3]. On August 21, 2008, the FCC issued an order which stated that Comcast's network management was unreasonable and that Comcast must terminate the use of its discriminatory network management by the end of the year. In December 2009 Comcast admitted no wrongdoing in its proposed settlement of up to 16 million dollars and decided to appeal the FCC's ruling with the US DC Court of Appeals to protect its rights, claiming that the FCC's decision was not based on any existing legal standards. The Court of Appeals sided with Comcast in its April 7, 2010 decision, pointing out that FCC has never been given by the Congress the authority to control the Internet network management, and agreed that there were no legal grounds for making Comcast stop its practice [4]. The court's decision means that broadband service providers are now free to manage their networks as they wish and this could prompt more discrimination on the Internet.

Net neutrality was first proposed by Columbia Law School professor Tim Wu, and is used to signify the idea that "maximally useful public information network aspires to treat all content, sites and platforms equally" [5]. While a formal process for the implementation of the principle does not exist, net neutrality usually means that broadband service providers charge consumers only once for the Internet access

© Springer Nature Switzerland AG 2020
P. Verma, F. Zhang, *The Economics of Telecommunication Services*, Textbooks in Telecommunication Engineering, https://doi.org/10.1007/978-3-030-33865-7_16

and do not favor one content provider over another, and do not charge content providers for sending varying amounts of information over broadband lines to end users [6]. Simply put, the network neutrality principle is that all Internet traffic should be treated equally, a philosophy which network neutrality proponents (online content providers like Google, Microsoft, and others) claim would preserve the principles on which the Internet was founded. Tim Berners-Lee, the founder of the World Wide Web, also favors keeping net neutrality in place, since "the Internet is the basis of a fair competitive market economy" [7].

However, Broadband service providers, like at&t, Verizon, and Comcast, among others, view net neutrality as being unfair to: (a) broadband service providers themselves and (b) light network users (compared to heavy users with the same access charge). Opponents also believe that prioritization of bandwidth is necessary for future innovation on the Internet [8]. The broadband service providers argue it is the service providers who have put their resources which they have to maintain and upgrade for their customers. They also argue that heavy-duty users and popular content providers (like Google, Skype) get a relatively "free ride" on their network which costs billions of dollars to build [9]. Lack of additional sources of revenue might act as a disincentive for broadband service providers to upgrade their infrastructure which, in turn, will affect the service providers' plans of increasing capacities. Further, it is estimated that 80% of Internet traffic is caused by 5% of users. This 5% generates all the traffic using P2P applications. In order to keep network traffic flowing for all consumers, broadband service providers argue it is reasonable for them to use network traffic management practices to slow down P2P application performances.

Broadband service providers claim that, net neutrality reduces the incentive to expand the capacity and the functionalities associated with the existing infrastructure for the next generation of broadband services. Conversely, when they are allowed to charge the online content providers for preferential treatment, the incentive is higher. Study [10], however, shows that the incentive for the broadband service provider to expand its services under net neutrality is unambiguously higher than under the no net neutrality regime. This is against the position of the broadband service providers that net neutrality results in limited incentive to expand. Results presented in [11] also indicate that non net neutrality networks are not always more favorable in terms of social welfare compared to net neutrality networks.

Although it seems reasonable for broadband service providers to choke certain applications which overwhelm the network in order for traffic flowing for everyone, people are afraid that broadband service providers will slow down or even block services and applications, they consider undesirable, freely. Especially under the situation that broadband and content providers are merging and turning to digital content distribution (e.g., Comcast bought NBC Universal on Jan, 2010), without net neutrality, as Lawrence Lessig and Robert W. McChesney said, "the Internet would start to look like cable TV: A handful of massive companies would be controlling access and distribution of content, deciding what you get to see and how much it would cost" [12]. In this chapter, we propose a solution for broadband service providers to control network congestion and maintain fairness among all consumers.

We propose the concept of inter-user compensations among users based on their usage of network resource. Users consuming less network resource will receive compensations from other users. One notable characteristic of this scheme is that the algebraic sum of all inter-user compensations is equal to zero which means that no inter-user compensations are lost. While compensations are among the users, broadband service providers' revenue are not affected by this scheme. In other words, broadband service providers' revenue remains constant under this model. We view this characteristic as being important since the network users would see such a mechanism positively in as much as the network provider does not take advantage of them by increasing its own profitability.

In this chapter, we assume that all users are responsible for the cost of the network and we model this cost sharing problem as a cooperative game. The cost share associated with each user corresponds to its Shapley Value of the cooperative game. We consider that the total cost of a coalition as the network resource required to maintain the desired QoS for the traffic in the coalition. The inter-user compensations are established based on the difference between their cost share and the actual price they pay by way of access fees to the broadband service provider. The broadband service provider can use a scale parameter to these inter-user compensations. During the peak period, the broadband service providers can increase these inter-user compensations to regulate the heavy users so as to control network congestion; otherwise, the broadband service provider can reduce these inter-user compensations to keep the network load at a desirable level. Another characteristic of this scheme is that the broadband service providers control network congestion and maintain fairness among all users without discriminatory treatment of any traffic flowing on the network. In other words, the net neutrality is well-maintained.

The rest of chapter is organized as follows. Section 16.2 presents the model used. Section 16.2 also includes the cooperative game and cost share. Section 16.3 investigates the inter-user compensations scheme. We present an illustrative example in Sect. 16.4. We also discuss the application of this mechanism. Section 16.5 captures our conclusions.

16.2   The Model

The network is modeled as a queuing network with First in First out (FIFO) discipline. Users in the network are denoted by a set as $N = \{1, \ldots, n\}$, and we use $|N|$ to denote the length of the set or the number of users in the network. The most prevalent pricing scheme in communication networks is access-rate dependent flat rate charge. For example, Verizon offers DSL Internet service which includes a starter package ($19.99 per month for download speeds up to 1.0 Mbps), power package ($29.99 per month for download speeds up to 3.0 Mbps), and turbo package ($39.99 per month for download speeds up to 10.0 Mbps). To emphasize our inter-user compensations mechanism, we further assume that *all users choose the same package.* Our mechanism can be easily extended to multiple packages of service when we consider inter-user compensations within the subscribers of each package.

Although all users pay the same amount to access the Internet, the traffic generated by each user is different. We use the average arrival rate λ_i to denote user i's traffic, and all users' traffic is assumed to follow Poisson distribution. The average delay D is used as the predefined QoS objective which is agreed to by network users and the network provider in their service level agreement (SLA). We also assume that the packet lengths of all packets in the network are exponentially distributed with the average packet length of $\frac{1}{\mu}$.

Based on [13], the aggregate traffic from a number of independent users with Poisson distribution still follows the Poissonian arrival discipline. We use λ to denote the aggregate traffic arrival rate. To maintain the required QoS D as described in the SLA, the required network resource c can be calculated using the M/M/1 queuing model as follows:

$$D = \frac{1}{\mu c - \lambda} \tag{16.1}$$

We can rewrite Eq. (16.1) as

$$c = \frac{1}{D\mu} + \frac{\lambda}{\mu} \tag{16.2}$$

From Eq. (16.2), we can find the total resource required to support the QoS promised by the network provider for the aggregate traffic λ. This problem of resource allocation among users generating varying levels of traffic can be modeled as a joint-cost allocation problem and we can use the Shapley Value to solve it.

The Shapley Value has been used as a method of joint-cost allocation instead of the traditional accounting allocation bases since the 1970s [14]. The Shapley Value was introduced by Shapley in 1953 as a method for each player to assess the benefits he or she would expect from playing a game. It consistently produces a unique allocation that virtually all researchers consider fair and equitable. The Shapely value method attributes the cost of a participant to the incremental cost associated with that participant. The Shapely value considers all orderings equally likely. This neutralizes any impact of the ordering in which the participants join the coalition. The allocation solution is thus fair [15].

To show its application to the problem of equitably assigning joint cost among individual users, the total network resource c as described in Eq. (16.2) among all users, we first define the worth of a coalition function $c(S)$. After that, we will compute the Shapley value. The coalition function $c(S)$ describes the required network resource to maintain the QoS for users in the coalition S, $S \subseteq N$ when the users in S share the network resource together. By (16.2), we find

$$c(S) = \frac{1}{D\mu} + \frac{\sum_{i \in S} \lambda_i}{\mu} \tag{16.3}$$

The incremental network resource used by a user $i \in N$ to the coalition S, $i \notin S$ is

$$c(S \cup \{i\}) - c(S) = \frac{\lambda_i}{\mu} \tag{16.4}$$

From Eq. (16.2) and Eq. (16.4) together, we can see the economy of scale in sharing the network resource: To maintain the average delay D, the incremental network resource required $\frac{\lambda_i}{\mu}$ for user i if users in $S \cup \{i\}$ sharing the resource together is smaller than the required network resource $\frac{1}{D\mu} + \frac{\lambda_i}{\mu}$ when user i was to be allocated the network resource individually.

The Shapley Value is defined as a weighted sum of the user's marginal contribution to all possible coalitions [16]. For all $i \in N$, the Shapley Value to user i is given by

$$SV_i = \sum_{S \subseteq N \setminus \{i\}} \frac{|S|!(|N| - |S| - 1)!}{|N|!} [c(S \cup \{i\}) - c(S)] \tag{16.5}$$

We note that, the term $c(S \cup \{i\}) - c(S)$ is the increase in network resource associated with user i in the coalition S. This incremental resource allocation occurs for those orderings in which the participant i is preceded by $|S|$ other players in $S \cup \{i\}$ and followed by $|N| - |S| - 1$ players not in $|S|$. This indicates that there are exactly $|S|!(|N| - |S| - 1)!$ orderings of interest. $|N|!$ is the total number of coalition permutations that can be created from the participants. Taken together, the expression $\frac{|S|!(|N| - |S| - 1)!}{|N|!}$ is the weighting factor that assigns equal share of the marginal contribution generated to each coalition of interest. The network resource consumed by the user i is thus weighted and aggregated for all possible coalitions where $i \in S$. Thus, user i is allocated a value equal to its expected incremental contribution in all possible coalitions. We have thus developed the desirable network resource share for each class by using the corresponding Shapley value.

Take Eq. (16.4) into Eq. (16.5), we get a more clear expression of network resource share for each user:

$$SV_i = \frac{1}{|N|D\mu} + \frac{\lambda_i}{\mu} \tag{16.6}$$

From Eq. (16.6), the network resource attributed to each user i depends on their usage λ_i: the larger the λ_i, the more the network resource consumed by user i. In addition, Eq. (16.6) also shows the economy of scale in the network resource consumption. For example, if user j's traffic is equal to $\lambda_j = 2 * \lambda_i$, the network resource share for user j is smaller than $2 * SV_i$; that is, $SV_j = \frac{1}{|N|D\mu} + \frac{\lambda_j}{\mu} < \frac{2}{|N|D\mu} + \frac{2 * \lambda_i}{\mu}$.

16.3   Inter-user Compensations Scheme

From Sect. 16.2, we find out how much network resource is consumed by each user using the Shapley Value in a joint-cost allocation circumstance. In the beginning of Sect. 16.2, we also stated that all users pay the same amount to access the network. In order to solve the cross-subsidization between light and heavy users, we propose an Inter-user compensations scheme. Besides the fairness between light users and heavy users, broadband service providers can also use the inter-user compensations scheme to control the behavior of heavy users to further control congestion in the network, flatten the peak hours of usage, and thus maintain the QoS.

The rationale behind the flat rate access is that all users are supposed to consume the same amount of network resource. From Eq. (16.2), the total network resource required to maintain average delay D for all users is

$$c(N) = \frac{1}{D\mu} + \frac{\sum_{i=1}^{n} \lambda_i}{\mu} \tag{16.7}$$

Therefore, each user is assumed to consume a network resource equal to $\frac{c(N)}{|N|}$, that is,

$$\hat{c} = \frac{c(N)}{|N|} = \frac{1}{|N|D\mu} + \frac{\sum_{i=1}^{n} \lambda_i}{|N|\mu} \tag{16.8}$$

Further, we define the inter-user compensation t_i as the difference between actual network resource it consumed, SV_i and its allocated resource consumption $\hat{c}$:

$$t_i = k * (SV_i - \hat{c}) = k * \frac{|N| * \lambda_i - \sum_{j=1}^{n} \lambda_j}{|N|\mu} \tag{16.9}$$

where k is a scale parameter discussed below.

Equation (16.9) shows that when $SV_i \geq \hat{c}$, user i consumed more network resource than it paid for the service and therefore it will compensate others by an amount equal to t_i. And when $SV_i \leq \hat{c}$, user i consumed less network resource than it paid for the service and therefore it will receive a compensation equal to t_i from others. The scale parameter k is used to control the inter-compensation dynamically to further control network congestion. For example, when total traffic on the network is heavy, the broadband service provider can increase k to increase the compensation heavy users give to light users in order to control heavy users' usage. This mechanism maintains fairness among users and solves the network congestion problem on an equitable basis.

We can check the algebraic sum of inter-user compensations as follows:

$$\sum_{i=1}^{n} = k * \sum_{i=1}^{n} (SV_i - \hat{c}) = k * \sum_{i=1}^{n} \frac{|N| * \lambda_i - \sum_{j=1}^{n} \lambda_j}{|N|\mu} = 0 \tag{16.10}$$

Equation (16.10) shows that sum of inter-user compensations is equal to zero and no inter-user compensations are lost. Thus the inter-user compensation mechanism is strictly among users. Broadband service providers use a scale parameter k to control the actual amount by which the heavy users compensate the light users. The circulation of the compensation is among the users only and the broadband service provider is a neutral party in this mechanism. The proposed mechanism thus guarantees the neutrality of the network. The broadband service provider's revenue is the same as before the proposed inter-user compensation mechanism was instituted. The flat rate access fee remained unchanged.

When the required network resource $c(N)$ to maintain the QoS is smaller than a threshold $c_{threshold}$, the scale parameter k is set to 0. The threshold can be assumed as a trigger for invoking the inter-user compensations mechanism. Broadband service providers can set this trigger point by observing the network traffic pattern. For example, the threshold could be set at the point corresponding to the network utilization of 0.5. When the total network load is low, the relatively heavy users should not be punished for using the network resource, thus encouraging them to shift their load to low usage periods.

When the required network resource $c(N)$ to maintain the QoS exceeds the threshold $c_{threshold}$, it signals the broadband service provider to control the heavy users' usage by setting the scale parameter k appropriately. We can assume k to be a function of $c(N)$, the larger the $c(N)$, the larger the value of k. In considering setting the inter-user compensation among users, the broadband service provider also needs to ensure that the compensations light users receive should always be smaller than the price of access.

16.4   An Example and Further Discussion

In this section, we first present a numerical example to show the inter-user compensation mechanism proposed in this chapter. Further, we discuss the application of this mechanism in broadband networks.

We assume there are three users in the network and the QoS parameter D agreed in the SLA is equal to 0.05 s. The average length of packets in the network is equal to 1, that is, $\frac{1}{\mu} = 1$. The average arrival for each user is $\lambda_1 = 5$; $\lambda_2 = 15$; $\lambda_3 = 20$. Based on Eq. (16.7), the total network resource required to maintain average delay 0.05 s is equal to

$$c(1, 2, 3) = \frac{1}{D\mu} + \frac{\lambda_1 + \lambda_2 + \lambda_3}{\mu} = 60$$

The network resource share for each user can be calculated based on (16.6):

$$SV_1 = \frac{1}{3 * D\mu} + \frac{\lambda_1}{\mu} = 11.67$$

$$SV_2 = \frac{1}{3 * D\mu} + \frac{\lambda_2}{\mu} = 21.67$$

$$SV_3 = \frac{1}{3 * D\mu} + \frac{\lambda_3}{\mu} = 26.66$$

Therefore, using Eq. (16.9) we can find the inter-user compensation as follows:

$$t_1 = k * \left(SV_1 - \frac{c(1, 2, 3)}{3} \right) = -8.33 * k$$

$$t_2 = k * \left(SV_2 - \frac{c(1, 2, 3)}{3} \right) = 1.67 * k$$

$$t_3 = k * \left(SV_3 - \frac{c(1, 2, 3)}{3} \right) = 6.66 * k$$

From the above, we find that user one will receive a compensation of $8.33 * k$; while users two and three will compensate user one by $1.67 * k$ and $6.66 * k$, respectively. The total inter-user compensation is equal to 0.

When the total available network resource is large, let us say the network capacity is equal to 100, then network utilization at this moment is

$$\rho = \frac{\lambda_1 + \lambda_2 + \lambda_3}{\mu * 100} = 0.4$$

If the broadband network provider assumes this network utilization is relatively low, then they can set $k = 0$ without deterring more traffic from entering the network and penalizing heavy users when such users do not cause congestion.

However, if the network capacity is equal to 50, the network utilization becomes

$$\rho = \frac{\lambda_1 + \lambda_2 + \lambda_3}{\mu * 50} = 0.8$$

In this case, the broadband service provider will choose k (which is no longer zero) appropriately in order to restrain the usage of users two and three so as to reduce the network utilization to a desirable level.

One reason that Comcast argued in its appeal to FCC's ruling with the US DC Court of Appeals is that Comcast does not target traffic from specific applications. Instead it begins to throttle traffic from heavy users. However, this also violates the philosophy of net neutrality since all Internet traffic should be treated equally by the neutral network. The proposed inter-user compensation scheme proposes an acceptable economic concept to manage the network without trying to target any specific entity within the network.

Since, in the proposed scheme, the broadband service provider is neutral and the compensations are circulated among users, this mechanism should be much more acceptable by the users as the traffic management scheme. It will also appeal to broadband service providers. The service provider can set up the threshold point and also the actual amount that the heavy users will pay light users. The service provider can be guided by the traffic pattern on the network and its own strategy to manage the network. For example, the service provider can trigger the inter-user compensations early but the compensations can be relatively small, or trigger the inter-user compensations late with relatively larger amounts.

Implementation of this mechanism will inevitably add more work to the network management system. However, from Eq. (16.9), the computational complexity of finding the inter-user compensation is linear and it should be acceptable by the broadband service provider.

16.5 Summary

In this chapter, we have proposed an inter-user compensations mechanism for a broadband service provider to manage the network traffic while maintaining the philosophy of net neutrality. Users consuming less network resource will receive compensation from heavy users. All these compensations are among users only and the broadband service provider keeps neutral in the process. The computational complexity of calculating the inter-user compensations is linear and this mechanism should be acceptable for the broadband service provider's management system. The broadband service provider's revenue remains constant. The proposed mechanism discourages heavy users during peak periods thus flattening the network usage.

Problems

16.1 It is estimated that 80% of Internet traffic is caused by 5% of the network users. To solve this unfairness, inter-user compensation is proposed in this chapter. Under the proposed scheme, does broadband service providers' revenue change?

16.2 How does the proposed inter-user compensation scheme maintain net neutrality?

16.3 User A consumes twice amount traffic as user B. Under the scheme proposed in this chapter, the network resource share (Shapley value) of user A is twice as user B. Is this correct?

References

1. TorrentFreak, *Comcast Throttles BitTorrent Traffic, Seeding Impossible* (2007). http://torrentfreak.com/comcast-throttles-bittorrent-traffic-seeding-impossible/, Retrieved Mar, 2020
2. P. Eckersley, F. Lohmann, S. Schoen, *Packet Forgery by ISPs: A Report on the Comcast Affair* (2007). http://www.eff.org/files/eff_comcast_report.pdf
3. Fox News, *FCC to Probe Comcast Data Discrimination* (2008). https://www.foxnews.com/story/fcc-to-probe-alleged-comcast-data-discrimination, Retrieved Nov, 2010
4. *U.S. Court Curbs FCC Authority on Web Traffic.* https://www.nytimes.com/2010/04/07/technology/07net.html, Retrieved Nov, 2010
5. T. Wu, Network neutrality, broadband discrimination. J. Telecommun. High Technol. Law **2**, 141 (2003). https://doi.org/10.2139/ssrn.388863.SSRN388863
6. R. Hahn, S. Wallsten, *The Economics of Net Neutrality* (The Berkeley Economic Press Economists' Voice, Columbia, 2006)
7. T. Berner-Lee, *Neutrality on the Net.* http://dig.csail.mit.edu/breadcrumbs/node/132, Retrieved Nov, 2010
8. H.D. Jonathan, *Internet Law, BNA Books* (2007), p. 750. ISBN 1570186839, 9781570186837
9. P. Waldmeir, *The Net Neutrality Dogfight that is Shaking up Cyberspace* (Financial Times, New York, 2006)
10. H.K. Cheng, S. Bandyopadhyay, G. Hong, Debate on net neutrality: a policy perspective. Inf. Syst. Res. **22**(1), 60–82 (2008)
11. J. Musacchio, G. Schwartz, J. Walrand, A two-sided market analysis of provider investment incentives with an application to the net-neutrality issue. Rev. Netw. Econom. **8**(1), 22–39 (2009)
12. L. Lessig, R.W. McChesney, *No Tolls on The Internet, The Washington Post* (2006), p. A23
13. L. Kleinrock, Queuing systems, in *Computer Applications*, vol. 2 (Wiley, New York, 1976)
14. A.E. Roth, R.E. Verrecchia, The Shapley value as applied to cost allocation: a reinterpretation. J. Account. Res. **17**(1), 295–303 (1979)
15. F. Maniquet, A characterization of the Shapley value in queueing problems. J. Econ. Theory **109**, 90–103 (2003)
16. Y. Chun, *A Note on Maniquet's Characterization of the Shapley Value in Queueing Problems*, Working Paper (Rochester University, Rochester, 2004)

Chapter 17
A Two-step Quality of Service Provisioning in Multi-Class Networks

17.1 The Model

In a common user network, the overall network resource is generally not dedicated to a single class or a certain user but shared by multiple classes and all users. Uncontrolled consumption of network resources might result in lower profit for the service provider and reduced satisfaction to users since it will lead to deteriorated throughput and unacceptable level of delay. In order to maximize the network utilization while guaranteeing service levels for different classes as described in SLAs, bandwidth allocation and flow control are generally enforced.

Since class-based network architecture is stateless from a per flow perspective, flow control should be enforced on the edge of network and the network resource should be allocated on a per-class basis in the network core. Reference [1] has considered an edge router for metering, policing, and shaping the incoming flows before aggregation into a limited number of classes. Recently, there are increasing numbers of responsible applications which are able to adjust their transmission rate according to the network condition. Unlike [1], we assume all flows are responsible flows in this chapter. Responsible flows have also been considered in [2–4]. The well-known resource scheduling schemes used in the core network include priority queuing (PQ), Weighted Round Robin (WRR), and Class-Based Queuing (CBQ) [5, 6]. However, these schedulers all keep a static service weight regardless of the actual number of aggregations in each class. Although dynamic schedulers like Fair WRR [1] have been proposed, the change of service weight is done somewhat arbitrarily without explicit network performance objective.

In this chapter, we model the network resource allocation among different classes (inter-class) as a centralized optimization problem to maximize the social welfare which is defined as the sum of all user utilities. For flow control in each class (intra-class), we develop a distributed game theoretic framework to regulate the individual flow behavior, where each flow competes for resources within the class to maximize its own performance.

© Springer Nature Switzerland AG 2020
P. Verma, F. Zhang, *The Economics of Telecommunication Services*, Textbooks in Telecommunication Engineering, https://doi.org/10.1007/978-3-030-33865-7_17

Modeling the network resource allocation as a centralized maximization social benefit problem on the basis of the knowledge of user utility functions has been considered in [4, 7, 8]. The maximization approach used in [7] solves the resource allocation problem in the context of network providing best-effort service. This chapter considers the resource allocation problem in a multi-class network environment where each class has an explicit QoS guarantee. In addition, unlike the utility function used in [7] which is only related to the transmission rate, we take not only the transmission rate but also the QoS parameter into the consideration for defining the utility function. In order to solve the scalability problem, when the number of sources becomes large, [7] forms a distributed flow control algorithm using gradient ascent algorithm from optimization theory. Our approach to the resource allocation mechanism does not suffer from the same scalability problem since the network provider only supports a limited number of classes. And the intra-class flow control mechanism is enforced separately using the Nash arbitration/bargaining framework.

The remainder of this chapter is organized as follows. Section 17.2 describes the network structure considered in this chapter. We introduce a game theoretic framework to control the flow behavior. For each class, Nash Arbitration Scheme (NAS) is computed and the characteristics of flows are explored. Section 17.3 shows dynamic network resources allocation among different classes to maximize social welfare while keeping the QoS in each class at a predefined value. Admission control conditions are also discussed. Section 17.4 gives a numerical example of the proposed flow control and resource allocation architecture and Sect. 17.5 captures the conclusions of this chapter.

17.2   Game Theoretic Framework

We consider a packet switched network with m classes as shown in Fig. 17.1. We model the network as a queuing network following the First in First out (FIFO) discipline. The capacity of the network C is distributed among the m classes in order to maximize social benefits which will be discussed in Sect. 17.3. Within each class, n_i users (or flows) share the allocated C_i and compete to maximize their individual performance objectives (defined later).

D_i is used as the predefined QoS objective for class i which is agreed to by class i users and the network provider in their SLA. T_i is the actual QoS experienced by class i flows. We denote a flow j in class i as $flow_{ij}$: $\lambda_{ij} \geq \lambda_{ijm}$, $T_i \leq D_i$, where λ_{ij} describes this flow's transmission rate and λ_{ijm} is the minimum rate requirement. We also assume the packet length of all flows is exponentially distributed with average length equal to $1/\mu$. Obviously, we have the actual average delay T_i for class i as:

$$T_i = \frac{1}{\mu C_i - \lambda_i} \tag{17.1}$$

where $\lambda_i = \sum_{j=1}^{n} \lambda_{ij}$ as the total traffic in class i.

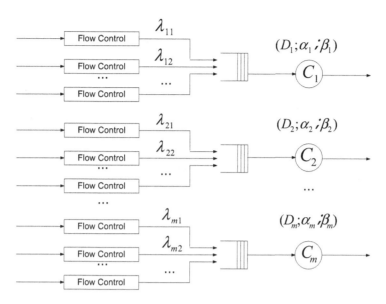

Fig. 17.1 Network model

Each flow is trying to maximize its performance objective, which is expressed as power given by Kleinrock in [9]: The weighted throughput of the system divided by the corresponding average delay in the system. We extend Kleinrock's definition to a multi-class situation. As a result, the power for flow j in class i is as follows:

$$P_{ij} = \frac{\lambda_{ij}^{\alpha_{ij}}}{T_i^{\beta_{ij}}} \tag{17.2}$$

where α_{ij}, β_{ij} are the sensitivity parameters for flow j in class i. α_{ij} describes the flow's sensitivity to the transmission rate and β_{ij} denotes the flow's tolerance to delay. We can see that if $\alpha_{ij} \geq \beta_{ij}$, the flow is more sensitive to transmission rate, or in another word, insensitive to delay, and vice versa. Without loss of generality, we assume $0 < \alpha_{ij} \leq 1, 0 < \beta_{ij} \leq 1$.

Normally, flows within the same class have the same QoS and minimum bandwidth requirement and, in our following analysis, we further assume flows belonging to the same class have identical sensitivity parameters and minimum transmission rate, that is $\alpha_i = \alpha_{ij}$, $\beta_i = \beta_{ij}$, $\lambda_{im} = \lambda_{ijm}$, $i = 1, \ldots, m; j = 1, \ldots, n_i$.

Within each class i, every flow competes for the limited allocated network resource C_i and tries to maximize its power. From Eq. (17.1) and Eq. (17.2), we can derive a more explicit description of power as follows:

$$P_{ij}(\lambda_i) = \lambda_{ij}^{\alpha_{ij}} * \left(\mu C_i - \sum_{j=1}^{n_i} \lambda_{ij}\right)^{\beta_{ij}} \tag{17.3}$$

As stated before, n_i is used to denote the number of flows in class i and vector $\lambda_i = (\lambda_{i1}, \lambda_{i2}, \ldots, \lambda_{in_i})$ is the transmission rate for each flow in class i. We can see from Eq. (17.3) that the power of *flow$_{ij}$* (P_{ij}) not only depends on its own transmission rate λ_{ij} but also on transmission rates of other flows in that class. Therefore, it is natural for us to model this problem as an n-party game.

The Pareto optimal point is a globally efficient solution and has better or at least equal payoff for each player than the Nash equilibrium point. We further assume that this game is cooperative and that the Pareto optimality can be found. Let a vector $\lambda_i^* = (\lambda_{i1}^*, \lambda_{i2}^*, \ldots, \lambda_{in_i}^*)$ be the transmission rate for each flow in class i under Pareto optimality. By the definition of Pareto optimality [10], the condition,

$$P_{ij}\left(\lambda_i^*\right) \leq P_{ij}\left(\lambda_i^* + \Delta\right), j = 1, \ldots, n_i \tag{17.4}$$

be met (Δ is a non-zero vector). It means that the Pareto optimality represents global maximization and it is impossible to find another point which leads to better payoff for at least one player without degrading the payoff to others.

In general, in a game with n players, the Pareto optimal points form an $n-1$ dimensional hypersurface and it means that there are an infinite number of points which are Pareto optimal. As we said before, an optimal network operation point is also a Pareto optimal point. The question here is: Which of these infinite Pareto optimal points should we choose for the network?

One way to find suitable Pareto optimal points for operation is by introducing further criteria. When we consider network resource sharing, one of the natural criteria is the notion of fairness. The notion of fairness is not well defined. There are many different ways to express it, like proportional fairness [7], max-min fairness [11], etc. In this chapter, we use the axioms of the fairness from game theory as the fairness criteria [12]. Nash arbitration scheme (NAS) which encapsulates the requirements of yielding Pareto optimality as well as satisfying the axioms of fairness is proposed in this section to find the suitable Pareto optimal point for each class. Stefanescu et al. [13] characterize the NAS as follows.

Let $f_j : X \to R, j = 1, 2, \ldots, n$ be concave upper-bounded functions defined on X which is a convex and compact subset of R^n and $f(x) = (f_1(x), \ldots, f_n(x))$.

Let $U = \{u \in R^n : \exists\ x \in X,\ \text{s.t.}\ f(x) \geq u\}$ and $X(u) = \{x \in X : f(x) \geq u\}$, $X_0 = X(u_0)$ be the subset of strategies that the users achieve at least their initial performance.

Then the NAS is given by the point which maximizes the following unique function:

$$\text{maximize}_x \prod_{j=1,\ldots,n} \left(f_j(x) - u_j^0\right) \tag{17.5}$$

From Eq. (17.3), $P_{ij}(\lambda_i)$ is defined on $(\lambda_{i1}, \lambda_{i2}, \ldots, \lambda_{in_i})$ and the transmission rate of each flow j in class i should be smaller than the service rate of each class i, μC_i, therefore, we have $0 \le \lambda_{ij} \le \mu C_i$, for $j = 1, \ldots, n_i$. Thus, $P_{ij}(\lambda_i)$ is defined on a convex and compact subset of R^{n_i}.

Now, we take the partial derivative and second partial derivative of $P_{ij}(\lambda_i)$ to check its concavity and upper-bound characteristics as follows:

$$\nabla_{\lambda_{ij}}(P_{ij}(\lambda_i)) = \lambda_{ij}^{\alpha_{ij}-1} \left(\mu C_i - \sum_{j=1}^{n_i} \lambda_{ij} \right)^{\beta_{ij}-1}$$
$$\times \left[\alpha_{ij} \left(\mu C_i - \sum_{k=1,k\neq j}^{n_i} \lambda_{ik} \right) - (\alpha_{ij} + \beta_{ij})\lambda_{ij} \right]$$

From above equation, we find that when $\lambda_{ij} \le \frac{\alpha_{ij}(\mu C_i - \sum_{k=1,k\neq j}^{n_i} \lambda_{ik})}{\alpha_{ij}+\beta_{ij}}$,

- P_{ij} is monotonically increasing with respect to λ_{ij}, and when $\lambda_{ij} \ge \frac{\alpha_{ij}(\mu C_i - \sum_{k=1,k\neq j}^{n_i} \lambda_{ik})}{\alpha_{ij}+\beta_{ij}}$,
- P_{ij} is monotonically decreasing with respect to λ_{ij}.

Therefore, we say that P_{ij} is upper-bounded at $\lambda_{ij} = \frac{\alpha_{ij}(\mu C_i - \sum_{k=1,k\neq j}^{n_i} \lambda_{ik})}{\alpha_{ij}+\beta_{ij}}$ given the $n_i - 1$ vector

$$(\lambda_{i1}, \ldots, \lambda_{i(k-1)}, \lambda_{i(k+1)}, \ldots, \lambda_{in_i}).$$

Now, let us check the second partial derivation of $P_{ij}(\lambda_i)$.

$$\nabla_{\lambda_{ij}}^2(P_{ij}(\lambda_i)) = \lambda_{ij}^{\alpha_{ij}-2} \left(\mu C_i - \sum_{j=1}^{n_i} \lambda_{ij} \right)^{\beta_{ij}-2} \left[\alpha_{ij}(\alpha_{ij}-1) \left(\mu C_i - \sum_{j=1}^{n_i} \lambda_{ij} \right) \right.$$
$$\left. + \beta_{ij}(\beta_{ij}-1) \left(\mu C_i - \sum_{j=1}^{n_i} \lambda_{ij} \right) - \alpha_{ij}\beta_{ij}\lambda_{ij}(\mu C_i - \sum_{j=1}^{n_i} \lambda_{ij}) \right]$$

From Eq. (17.1), we know $(\mu C_i - \sum_{j=1}^{n_i} \lambda_{ij}) \ge 0$ since delay cannot be negative. And we already assumed in Sect. 17.2 that the sensitive parameters α and β are defined as, $0 < \alpha \le 1$ and $0 < \beta \le 1$. From above equation, we conclude that $\nabla_{\lambda_{ij}}^2 P_{ij}(\lambda_i) < 0$ and P_{ij} is a concave function on λ_{ij}.

We also assume that each user has an initial arrival rate 0 and now we are ready to calculate the suitable Pareto optimality by solving the following maximization problem:

$$\text{maximize}_{\lambda_i} \prod_{j=1,\ldots,n_i} (P_{ij}(\lambda_i)) \tag{17.6}$$

which leads to

$$\nabla_{\lambda_{ij}} \left(\prod_{j=1,\ldots,n_i} (P_{ij}(\lambda_i)) \right) = 0, \; j = 1, \ldots, n_i \tag{17.7}$$

And Eq. (17.7) is equivalent to

$$\alpha_{ij} \left(\mu C_i - \sum_{j=1}^{n_i} \lambda_{ij} \right) - \lambda_{ij} \left(\sum_{j=1}^{n_i} \beta_{ij} \right) = 0, \; j = 1, \ldots, n_i \tag{17.8}$$

From Eq. (17.8), we obtain a linear system of equations with a unique solution as follows:

$$\lambda_{ij}^* = \frac{\mu C_i \alpha_{ij}}{\sum_{j=1}^{n_i} (\alpha_{ij} + \beta_{ij})} \tag{17.9}$$

Given the assumption in Sect. 17.2 that all flows within the same class have identical QoS requirement, the transmission rate for each flow in class i under NAS therefore is

$$\lambda_{i1}^* = \lambda_{i2}^* = \ldots = \lambda_{in_i}^* = \frac{\mu C_i}{n_i \left(1 + \frac{\beta_i}{\alpha_i} \right)} \tag{17.10}$$

The actual system delay experienced by each flow can be calculated using Eq. (17.1):

$$T_i = \frac{1 + \frac{\alpha_i}{\beta_i}}{\mu C_i} \tag{17.11}$$

Equation (17.10) shows that under NAS, each flow has identical transmission rate. In the definition of power in Sect. 17.2, α_i and β_i are related to the QoS requirements for each class. Equation (17.11) further exemplifies this statement in a way that the parameter T_i is dependent on $\frac{\alpha_i}{\beta_i}$ with a given class resource C_i. From Eq. (17.11), we also find that the equilibrium T_i does not degrade by the increasing number of flows in the class. Therefore, we can use $\frac{\beta_i}{\alpha_i}$ as a QoS indicator of class i. The bigger $\frac{\beta_i}{\alpha_i}$ is, the smaller T_i becomes and the better QoS class i gets.

Under the game theoretic flow control framework described in this section, the QoS of each class can be maintained no matter how many flows reside in them. With this property, a fair and efficient inter-class resource allocation mechanism among different classes is proposed in Sect. 17.3.

17.3 Inter-class Resource Allocation

The network resource is allocated dynamically among classes based on the network situation to maximize the social benefit. Social benefit is the sum of the utility functions of each user. The most well-known utility function as proposed by Kelly [7] has the form $U_{ij} = w_{ij} \log \lambda_{ij}$, where w_{ij} is user's willingness to pay and λ_{ij} is the transmission rate.

In this chapter, we define a new utility function as follows:

$$U_{ij} = \frac{\beta_i}{\alpha_i} \log \lambda_{ij} \tag{17.12}$$

Equation (17.12) describes that the utility function's relation to the QoS indicator $\frac{\beta_i}{\alpha_i}$ and the transmission rate λ_{ij}. This is consistent with the notion that the user's utility is proportional to the QoS and the logarithm of the transmission rate since the better the QoS, the more utility the user gets. This utility function also fits into Kelly's utility function in a way such that the better the QoS, the more willingness to pay for the service. The reason we use a logarithmic function lies in the marginal utility as a function of transmission rate diminishing as the rate increases. This is consistent with [14].

In Sect. 17.2, we have shown that under NAS in each class, each flow converges to the identical flow rate. As a result, the sum of all flows' utility function in each class i can be given as

$$U_i = n_i \frac{\beta_i}{\alpha_i} \log \lambda_{ij} = n_i \frac{\beta_i}{\alpha_i} \log \frac{\mu C_i}{n_i \left(1 + \frac{\beta_i}{\alpha_i}\right)} \tag{17.13}$$

As assumed in Sect. 17.2, there are m classes in the network, therefore the social benefits maximization problem becomes

$$\text{maximize}_{c_i} \sum_{i=1}^{m} n_i \frac{\beta_i}{\alpha_i} \log \lambda_{ij} = \sum_{i=1}^{m} n_i \frac{\beta_i}{\alpha_i} \log \frac{\mu C_i}{n_i \left(1 + \frac{\beta_i}{\alpha_i}\right)} \tag{17.14}$$

subject to the following constraints:

$$\sum_{i=1}^{m} C_i \leq C \tag{17.15}$$

$$T_i = \frac{1 + \frac{\alpha_i}{\beta_i}}{\mu C_i} \leq D_i, i = 1, \ldots, m \tag{17.16}$$

$$\lambda_i = \frac{\mu C_i}{1 + \frac{\beta_i}{\alpha_i}} \geq n_i \lambda_{im}, i = 1, \ldots, m \tag{17.17}$$

Inequality (17.15) states that the resource allocated among all m classes subject to the total available network resource C. Inequality (17.16) holds that for each class, the average delay experienced should be smaller than the promised parameter in SLA. It has been shown in Sect. 17.2 that, under NAS, each flow within the same class has the same transmission rate and Eq. (17.17) is used to maintain minimum bandwidth for each flow.

We simplify the constraints (17.16) and (17.17), then get

$$C_i \geq \frac{1 + \frac{\alpha_i}{\beta_i}}{\mu D_i}, i = 1, \ldots, m$$

and

$$C_i \geq \frac{n_i \left(1 + \frac{\beta_i}{\alpha_i}\right) \lambda_{im}}{\mu}, i = 1, \ldots, m$$

Thus, constraints (17.16) and (17.17) can be simplified as a single constraint as follows:

$$C_i \geq C_{im}, i = 1, \ldots, m \tag{17.18}$$

where $C_{im} = \max(\frac{1 + \frac{\alpha_i}{\beta_i}}{\mu D_i}, \frac{n_i(1 + \frac{\beta_i}{\alpha_i})\lambda_{im}}{\mu})$.

When $\sum_{i=1}^{m} C_{im} \geq C$, the network is oversubscribed and by no means can provide the promised QoS and minimum bandwidth requirement for each flow as defined in SLA, and therefore admission control has to be introduced. As a result, for the admission control, we have

$$\sum_{i=1}^{m} C_{im} \leq C \tag{17.19}$$

It can be observed that the expression in Eq. (17.14) is a strictly concave function over a closed and bounded set defined by Eq. (17.15) and Eq. (17.18). Therefore a unique maximum always exists. We use Lagrangian multipliers and get

$$\text{maximize}_{C_i} \sum_{i=1}^{m} n_i \frac{\beta_i}{\alpha_i} \log \frac{\mu C_i}{n_i \left(1 + \frac{\beta_i}{\alpha_i}\right)} - \gamma_0 \left(\sum_{i=1}^{m} C_i - C\right) + \sum_{i=1}^{m} \gamma_i \left(C_i - C_{im}\right)$$

$$(17.20)$$

The necessary and sufficient Karush–Kuhn–Tucker (KKT) [15] conditions applicable to Eq. (17.20) are given by

$$\frac{n_i \beta_i}{C_i \alpha_i} - \gamma_0 + \gamma_i = 0 \qquad (17.21)$$

$$\gamma_0 \left(\sum_{i=1}^{m} C_i - C\right) = 0 \qquad (17.22)$$

$$\gamma_i (C_i - C_{im}) = 0, \, i = 1, \ldots, m \qquad (17.23)$$

We denote the optimum network resource allocation among m classes as $(C_1^*, C_2^*, \ldots, C_m^*)$.

This centralized network resource allocation mechanism does not suffer from the scalability problem for the reason that the network provider only needs to maintain a limited number of classes of service and our resource allocation is made among these classes. Furthermore, since the utility function we used in this section is logarithmic, the solution obtained has the proportionally fair property as shown by Kelly in [7].

Thus, the resource allocation state in this chapter is an efficient and fair mechanism from both intra-class and inter-class perspectives. The intra-class resource allocation is modeled as a multi-party game. Each flow tries to maximize its power in a distributed way. On the NAS, each flow has identical fraction of network resources and the QoS is not impacted by the number of flows in the class. The inter-class resource allocation is based on maximizing social benefits while allowing each class to maintain its QoS and minimum bandwidth requirement.

We can now make some observations regarding the admission control in each class.

In class i, when the allocated resource C_i^* satisfies the following equation:

$$\frac{\mu C_i^*}{1 + \frac{\beta_i}{\alpha_i}} = n_i \lambda_{im} \qquad (17.24)$$

the admission control is considered to begin in this class. Equation (17.24) suggests that using the KKT condition (17.23), $C_i^* - C_{im} = 0$. And together with Eq. (17.18), we have

$$T_i^* = \frac{1 + \frac{\alpha_i}{\beta_i}}{\mu C_i^*} \leq D_i \tag{17.25}$$

From Eq. (17.11) and Eq. (17.24), the average delay for class i can be rewritten as

$$T_i^* = \frac{\alpha_i}{\beta_i n_i \lambda_{im}} \tag{17.26}$$

Equations (17.25) and (17.26) show that in class i, the flows will receive the lower delay T_i^* than the SLA agreed delay D_i and T_i^* decreases with increasing number of flows in this class for the scale efficiency. Admission control is considered when (17.24) is first satisfied. The reason is that the network load is becoming heavy when Eq. (17.24) is met. Another reason is that since T_i^* is already lower than D_i, decreasing T_i^* is not as valuable as the increasing the transmission rate for flows in other classes. Therefore, this is the point at which admission control for that class needs to be introduced.

17.4   An Illustrative Example

This section presents a numerical example about the network resource allocation and flow control mechanisms proposed in this chapter. We will develop the QoS of each class and allocated bandwidth of each flow profile using the proposed mechanisms.

For simplicity, we assume the network has three different service classes ($m = 3$) with the capacity $C = 100$. The service class 1 is supposed to support real-time gaming and the QoS parameter average delay defined in SLA as $D_1 = 0.04$s; while each flow within this class has the identical bandwidth to delay weighting factor $\frac{\alpha_1}{\beta_1} = 0.9$ and minimum transmission rate requirement $\lambda_{1m} = 1.8$. Interactive streaming service will be carried in class 2; the delay and transmission rate requirement are as $D_2 = 0.1$s, $\lambda_{2m} = 1.2$, the bandwidth to delay sensitive parameter as $\frac{\alpha_2}{\beta_2} = 1.1$. The service class 3 is designed to support non-interactive streaming service and has following parameters: $D_3 = 0.3\,s$, $\lambda_{3m} = 0.8$, $\frac{\alpha_3}{\beta_3} = 1.5$. We also assume the average length of packets in the network $\frac{1}{\mu} = 1$.

Table 17.1 shows the network resource allocation among the three classes using the mechanisms proposed in this chapter under different network scenarios.

For example, in scenario 1 when there are 6 flows in class 1, 5 flows in class 2, and 4 flows in class 3, class 1 will be allocated with 48.0%, class 2 will be allocated with 32.8%, and class 3 with 19.2% of the network resource.

Under the network resource allocation shown in Table 17.1, using the flow control mechanism described in Sect. 17.2, Fig. 17.2 describes the average delay curve of each class under different network scenarios shown in Table 17.1 and

Table 17.1 Network resource allocation under different network scenarios

	Scenario 1	Scenario 2	Scenario 3	Scenario 4	Scenario 5	Scenario 6	Scenario 7	Scenario 8	Scenario 9
Number of flows in class 1	$n_1 = 6$	$n_1 = 7$	$n_1 = 8$	$n_1 = 9$	$n_1 = 10$	$n_1 = 11$	$n_1 = 12$	$n_1 = 13$	$n_1 = 14$
Number of flows in class 2	$n_2 = 5$	$n_2 = 6$	$n_2 = 7$	$n_2 = 8$	$n_2 = 9$	$n_2 = 10$	$n_2 = 11$	$n_2 = 12$	$n_2 = 13$
Number of flows in class 3	$n_3 = 4$	$n_3 = 5$	$n_2 = 6$	$n_3 = 7$	$n_3 = 8$	$n_3 = 9$	$n_3 = 10$	$n_3 = 11$	$n_3 = 12$
Allocated resource for class 1	$C_1 = 48.0$	$C_1 = 47.5$	$C_1 = 47.5$	$C_1 = 47.5$	$C_1 = 47.5$	$C_1 = 47.5$	$C_1 = 47.5$	$C_1 = 49.4$	$C_1 = 53.2$
Allocated resource for class 2	$C_2 = 32.8$	$C_2 = 32.7$	$C_2 = 32.2$	$C_2 = 32.0$	$C_2 = 31.8$	$C_2 = 31.6$	$C_2 = 31.5$	$C_2 = 30.3$	$C_2 = 29.8$
Allocated resource for class 3	$C_3 = 19.2$	$C_3 = 19.8$	$C_3 = 20.3$	$C_3 = 20.5$	$C_3 = 20.7$	$C_3 = 20.9$	$C_3 = 21.0$	$C_3 = 20.3$	$C_3 = 17.0$

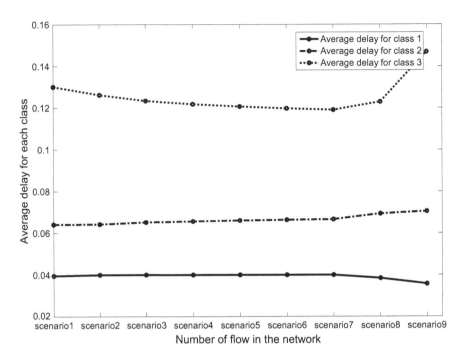

Fig. 17.2 Delay experienced by each class under different network scenarios

Fig. 17.3 presents how the flow transmission rate in each class changes with the increasing number of flows in the network.

As can be seen from Fig. 17.2, even as the average delay of each class changes, they all satisfy the SLA QoS agreement; the average delay of class 1 is less than 0.04 s; the average delay of class 2 is less than 0.1 s; the average delay for class 3 is less than 0.3 s. In Fig. 17.3, it can be observed that although the flow rate of each class decreases as the number of flows increases, they all satisfy the minimum transmission rate requirement in each class. The minimum flow rate in class 1 is 1.8; minimum flow rate in class 2 is 1.2, and minimum flow rate in class 3 is 0.8.

In Fig. 17.3, the flow rate for class 1 is kept at the minimum rate 1.8 in scenarios 8 and 9. In Fig. 17.2, we find that the average delay for class 1 in scenario 8 and 9 is decreasing. This gives a signal to consider admission control in class 1 as described in Sect. 17.3.

Figure 17.3 also shows the proportionally fair transmission rate of the resource allocation mechanisms described in this chapter. The network resources allocation mechanism does not favor any special class but allocates the resources in a way to maximize the social benefits. For example, in scenario 1, the flow rate in class 1 is 3.8, flow rate in class 2 is 3.4, and flow rate is class 3 is 2.8. All these flow rates are above their minimum requirement by different percentages and the network utility is maximized.

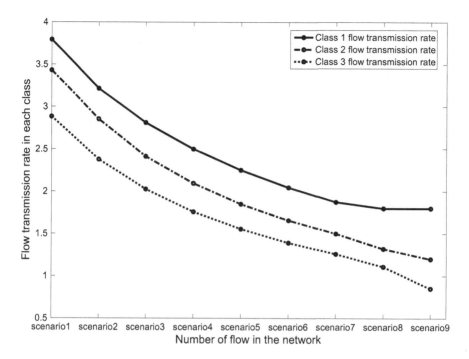

Fig. 17.3 Flow rate in each class under different network scenarios

17.5 Summary

In this chapter, we have presented a two-step mechanism to provide QoS in the multi-class network environment. The inter-class resource allocation problem is modeled as a centralized optimization problem to maximize the sum of all users' utilities where we define the utility as a function of each user's transmission rate and the QoS it received. This centralized optimization does not suffer from the scalability problem since the network only needs to maintain a limited number of classes. We use a game theoretic framework to control the optimal rate within each class leading to the NAS expected in a distributed manner. As shown in Sect. 17.4, these mechanisms assure the QoS for each flow as described in the SLA. Further, the resource allocation to each flow results in maximizing the social benefits of the network.

Problems

17.1 Using the model proposed in this chapter, how do we allocate network resource among different classes? How does the model allocate resource to each flow within a class?

17.2 In this chapter, the Nash Arbitration Scheme (NAS) is used to find the Pareto optimal point within each class. Under NAS, what is the transmission rate for each flow?

References

1. S. Yi, X. Deng, G. Kesidis, C.R. Das, Providing fairness in DiffServ architecture, in *Global Telecommunications Conference*, vol. 2 (2002), pp. 1435–1439
2. X. Wang, H. Schulzrinne, Pricing network resources for adaptive applications. IEEE/ACM Trans. Netw. **14**(3), 506–519 (2006)
3. S. Shenker, Fundamental design issues for the future. Int. IEEE J Sel. Areas Commun. **13**, 1176–1188 (1995)
4. F. Kelly, A. Maulloo, D. Tan, Rate control in communication networks: shadow prices, proportional fairness and stability. J. Oper.Res. Soc. **49**, 237–252 (1998)
5. V. Jacobson, K. Nichols, K. Poduri, *An Expedited Forwarding PHB Group, IETF RFC 2598* (1999)
6. G. Mamais, M. Markaki, G. Politis, I.S. Venieris, Efficient buffer management and scheduling in a combined IntServ and DiffServ architecture: a performance study, in *Proceedings of the ICATM* (1999), pp. 236–242
7. F.P. Kelly, Charging and rate control for elastic traffic. Eur. Trans. Commun. **8**, 33–37 (1997)
8. A. Ganesh, K. Laevens, R. Steinberg, Congestion pricing and user adaptation, in *Proceedings of the IEEE INFOCOM, Anchorage, AK* (2001), pp. 959–965
9. L. Kleinrock, Power and deterministic rules of thumb for probabilistic problems in computer communications, in *International Conference on Communications, Boston, Mass* (1979), pp. 43.1.1–43.1.10
10. T. Basar, G.J. Oldser, *Dynamic Noncooperative Game Theory* (Academic Press, New York, 1982)
11. D. Bertsekas, R. Gallager, *Data Networks* (Prentice-Hall, Inc., Upper Saddle River, 1987)
12. J. Nash, The bargaining problem. Econometrica **18**, 155–162 (1950)
13. A. Stefanescu, M.W. Stefanescu, The arbitrated solution for multiobjective convex programming. Rev. Roum. Math. Pure Applicat. **29**, 593–598 (1984)
14. A. Watson, M.A. Sasse, Evaluating audio and video quality in low-cost multimedia conferencing systems. Interact. Comput. **8**(3), 255–275 (1996)
15. D.P. Bertsekas, *Nonlinear Programming* (Athena Scientific, Belmont, 1999)

Chapter 18
Network of the Future

18.1 Need for Advanced Level of Control

Experiential evidence gathered over the past decade or more has shown enormous benefits associated with having minimal levels of control, especially at the global level in networks. We have seen that a network composed of relatively autonomous nodes acting on the basis of information gathered from nodes in the immediate vicinity has become a principal target for network designers. The benefits of having global control in the network appear to be small in relation to the cost of gathering, analyzing, and acting upon network-wide data. The general perception is that a loosely controlled network can more easily accommodate network expansion on an ad-hoc basis and can be more easily realized using potentially divergent architectures.

There is little doubt that the future common user network will be required to offer differentiated levels of service at varying prices. Several chapters in the book have examined the need for and the consequences of such a policy and how it can be offered using game theory where both the offeror and the receivers of differentiated services can reach a level of equilibrium.

Video services have likely already become the dominant part of the Internet traffic. Pricing for video traffic would sky rocket if the video traffic had the same level of quality of service requirement (in terms of delay and jitter) as interactive human voice service. (This is more fully discussed in the next section.) During high levels of traffic because of the Poissonian characteristics of telecommunication traffic, the network must find a way to cope with the excessive demand and continue to perform at its rated capacity. There are only two mechanisms available for delivering acceptable levels of quality of service in a network subject to statistical variations in incident traffic: Inject additional bandwidth resources at the points of congestion, or throttle traffic right at the point of origin before it enters the network. Either of these techniques requires continually updated information on the network

© Springer Nature Switzerland AG 2020

P. Verma, F. Zhang, *The Economics of Telecommunication Services*, Textbooks in Telecommunication Engineering, https://doi.org/10.1007/978-3-030-33865-7_18

at the link, nodal, and global levels. It leads us to conclude that the control plane of the network of the future must have visibility of the network as a whole.

18.2   Pricing of Network Services

Pricing of network services is a complex issue and is driven by regulatory and competitive factors in addition to a corporation's strategy and the underlying cost structure of delivering services. On a short run, the impact of competitive or regulatory elements might be overwhelming considerations for determining price. However, the main thesis of this book is that the only sustainable basis of pricing is based on revenue and the cost of producing that revenue.

Communication services, even commodity communication services, have a very important and differentiating characteristic. Unlike other services, e.g., water distribution, power supplies from an electric grid, delivery of healthcare, etc., the consumption of resources in a communication network is constant and is independent of the volume of incident traffic or the served traffic. It is important, therefore, to maximize the utilization of the resources the network has invested in. This can be achieved by insuring that the throughput of the network or the served traffic does not fall under excessive load and continues to perform at its rated capacity. Several chapters in the book have shown how this can be realized.

The pricing model used in the past by carriers has been based on the level of resources consumed by the traffic being priced. The pricing model we have proposed in this book is based on the cost of displaced opportunity as opposed to the cost of the elements of the network engaged in delivering the service. The displaced opportunity is characterized by the displaced service that the network could have alternatively delivered more efficiently. A price for the most efficient service that a network can deliver is determined based on the profitability target for the network as a whole. Once this is known, pricing for other, possibly higher-function services, can be derived on the basis of imputed revenue that the most efficient service would have generated using the same level of resources. This would provide a robust base on which sustainable pricing models for a range of differentiated services offered by a single common user network can be developed.

An example of pricing differentiated services with different qualities of service offered off a single network fabric might be in order. Video transport services, more specifically, the transfer of a video stream from one buffer to another, can withstand a level of variable delay that will be unacceptable for human conversation. Since a video service can withstand a large level of jitter, it would have a relatively lower level of bandwidth requirement in a common user network. The reason for this discrepancy is based on the queuing characteristics of telecommunication traffic. The telecommunication traffic is most generally characterized by Poisson arrivals and negative exponential service time distribution. The negative exponential distribution has a very important characteristic: the variance of the distribution is equal to the mean. If the jitter (measured as variance of delay) offered to video streaming traffic were to be identical to the jitter associated with interactive traffic such as human conversation, the needed bandwidth for video traffic would be very large indeed. By the same token, if the tolerable limit of jitter for video traffic were much higher, the need for bandwidth would be much lower. In that case, the transfer of a video streaming bit would need a lower level of network resources than the bit which is part of interactive traffic with a low jitter requirement. Video streaming services can thus be priced at a much lower level than traffic necessitating low delay and low jitter.

18.3 The Network as a Surrogate for Privacy and Security

What level of privacy and security can a network offer? Is it desirable for a network to remain simply as the carrier without any interest in, or responsibility for the consequences of, the payload it conveys between and among the communicating parties it connects? We do not know the answer to the latter question but offer some possibilities for the former.

18.3.1 Network-Based Security

There are three primary properties of secure communication: Confidentiality, authentication, and integrity. In this sub-section, we only consider the authentication function since the other two can be accomplished by well-known techniques in cryptography that can be realized at data speeds.

For real-time verbal communication among known individuals, authentication is guaranteed by the innate abilities of speaker recognition. This possibility is somewhat compromised by the fact that authentication might need to precede actual communication, although this limitation can be alleviated by a voice recognition system that might ask a caller to speak her name or produce some authentication function mutually agreed to between the network and the communication initiating party.

The general task of authentication might be relegated to the end points as in the preceding paragraph. A network provider might, however, like to positively identify and log the party that initiates a call. In the legacy PSTN network, the calling line identification (CLID) served this purpose. The CLID was tied to a physical address where the call originating telephone was located. In the PSTN, the identity of the called party was similarly known to the caller. The PSTN thus offered a network-based authentication mechanism.

If a network were to perform the identification function, there would be at most n functions needed for a base of n users. On the other hand, if the identification function were to be performed by the end points, $n(n-1)$ possible functions would be needed to be performed. The need for this quadratic increase might be obviated by a network-based authentication mechanism. A higher speed of operation resulting from a reduction in computational overhead and key management issues might be expected as a result.

A requirement for a network-based authentication is the existence of the network as a trusted agent of the end point. Depending on the called party's preference, however, the network might make the identity of the calling party available to the called party. The called party can refuse to accept communication from a party the network cannot positively identify by so communicating its preference to the network on an ad-hoc basis. Calls across networks would need to perform network-to-network based identification functions in order to make the integrity of identification seamless across networks. The calling party can, reciprocally, require the network to positively identify the called party unless the latter has opted out. In this case, the call cannot be completed.

18.4   Summary

This chapter has examined the likely structure of future networks. It has pointed out the need for control at the global level in future networks in order to effectively deliver services with widely varying traffic characteristics and performance requirements. Furthermore, it has concluded that a network can function as a reliable surrogate for implementing privacy and security functions for its clients.

Appendix A
Show that $\frac{\partial \ln n}{\partial n} = \frac{1}{n}$

Proof For any function $f(x)$,

$$f'(x) = \lim_{h \to 0} \frac{f(x+h) - f(x)}{h}, \text{ therefore,}$$

$$\frac{\partial \ln n}{\partial n} = \lim_{h \to 0} \frac{\ln(n+h) - \ln(n)}{h}$$

$$= \lim_{h \to 0} \frac{1}{h} \ln\left(\frac{n+h}{n}\right)$$

$$= \lim_{h \to 0} \frac{1}{n} \frac{n}{h} \ln\left(\frac{n+h}{n}\right)$$

$$= \frac{1}{n} \lim_{h \to 0} \frac{n}{h} \ln\left(\frac{n+h}{n}\right)$$

$$= \frac{1}{n} \lim_{h \to 0} \ln\left(1 + \frac{h}{n}\right)^{\frac{n}{h}}$$

$$= \frac{1}{n} \ln e$$

$$= \frac{1}{n}$$

which proves the result.

© Springer Nature Switzerland AG 2020
P. Verma, F. Zhang, *The Economics of Telecommunication Services*, Textbooks in
Telecommunication Engineering, https://doi.org/10.1007/978-3-030-33865-7

Appendix B
Merger of Heterogeneous Networks

Underlying the motivation for merger of two separate networks discussed in Chap. 1 was the fact that each user presented a uniform demand on the network and contributed an identical value to the network. This appendix examines the case for merger of, or separation into, two or more networks as merged or separate entities, to carry traffic that differ in characteristics [1]. Is merging such networks always desirable as was the case when each user presented the same demand and the associated value?

In this analysis, we assume the switching technology used by each network is based on packet switching, which is the technology on which the Internet is based. In particular, we ask the following question: If the total resources used by two different networks were held to a constant value, would it be better to assign the total resource to a combined network, or would it be better to split the resources in some optimal fashion and assign them to two different networks, each carrying a separate class of traffic. The separation of traffic into two different classes is important because it can be easily proven that if there is a single class of traffic, a combined network will better serve its customers if the network resources were merged into a single network which carried the merged traffic.

In the following analysis, we assume that the network resource is bandwidth. The characterization of performance is in terms of average delay and the question, mathematically, is whether two different transmission systems each assigned to one specific class of traffic would perform better or worse than a combined system. We have simplified a general problem into a simpler construct in order to gain an insight into the benefit of integrating transmission resources for heterogeneous traffic.

Two classes of traffic are assumed. The length of message associated with each traffic class is geometrically distributed. The arrival for each class of traffic is characterized by Poisson distribution. Table B.1 presents the parameters of the two classes of traffic; traffic class 1 is served by System S_1 and traffic class 2 by System S_2. Similarly, Table B.2 presents the traffic parameters for the combined or the integrated system. The number of messages per second from each class is λ_1 and

© Springer Nature Switzerland AG 2020

P. Verma, F. Zhang, *The Economics of Telecommunication Services*, Textbooks in Telecommunication Engineering, https://doi.org/10.1007/978-3-030-33865-7

Table B.1 Segregated case

System	Average no. of messages	Mean length	Average message delay under optimum channel capacity	Capacity allocated
s_1	λ_1	$\frac{1}{\mu_1}$	T_1	c_1
s_2	λ_2	$\frac{1}{\mu_2}$	T_2	c_2

Table B.2 Integrated case

System	Average no. of messages	Mean length	Average message delay under optimum channel capacity	Capacity allocated
s	$\lambda_1 + \lambda_2$	$\frac{1}{\mu_1}\frac{\lambda_1}{\lambda_1+\lambda_2} + \frac{1}{\mu_2}\frac{\lambda_2}{\lambda_1+\lambda_2}$	T	$c_1 + c_2$

Fig. B.1 Segregated and integrated system

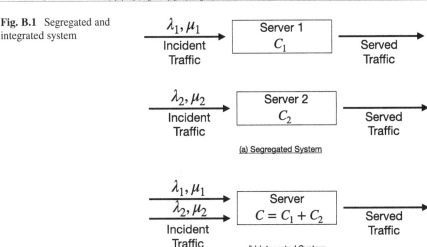

(a) Segregated System

(b) Integrated System

λ_2, respectively. S_1 is assigned a capacity of c_1 and S_2 is assigned a capacity of c_2. For the combined system S, the total number of messages per second is indicated in Table B.2. The combined capacity $c = c_1 + c_2$ is allocated to the System S.

Figure B.1a shows the segregated system and Fig. B.1b shows the integrated system.

It can be shown that [2], for a single-channel system with Poisson arrivals and an arbitrary distribution of message lengths, the expected time a message spends in the system is

$$T = \frac{1}{\mu c} + \frac{\rho^2 + \lambda^2 \sigma^2}{2\lambda(1 - \rho)} \tag{B.1}$$

where

- $\frac{1}{\mu}$ = mean message length (*bits*),
- c = channel capacity (*bits/s*),

- $\sigma^2 =$ variance of service time (s^2),
- $\lambda =$ arrival rate of message (s^{-1}),
- $\rho =$ utilization factor $= \frac{\lambda}{\mu c}$.

For the segregated system, we have [2]

$$\sigma^2 = \frac{1}{c^2} \frac{1 - \mu}{\mu^2} \tag{B.2}$$

leading to

$$T_1 = \frac{1}{2c_1} \frac{2c_1 - \lambda_1}{\mu_1 c_1 - \lambda_1} \tag{B.3}$$

and

$$T_2 = \frac{1}{2c_2} \frac{2c_2 - \lambda_2}{\mu_2 c_2 - \lambda_2} \tag{B.4}$$

Furthermore, for the segregated system, the mean delay per message T_{seg} averaged over the systems can be given as

$$T_{seg} = \frac{\lambda_1}{\lambda_1 + \lambda_2} T_1 + \frac{\lambda_2}{\lambda_1 + \lambda_2} T2 \tag{B.5}$$

or,

$$T_{seg} = \frac{\lambda_1}{\lambda_1 + \lambda_2} \frac{1}{2c_1} \frac{2c_1 - \lambda_1}{\mu_1 c_1 - \lambda_1} + \frac{\lambda_2}{\lambda_1 + \lambda_2} \frac{1}{2c_2} \frac{2c_2 - \lambda_2}{\mu_2 c_2 - \lambda_2} \tag{B.6}$$

Since the total capacity c is fixed, the minimum value of T_{seg} as c_1 (or c_2) is varied can be obtained by setting [2]

$$\frac{\partial T_{seg}}{\partial c_1} = 0 \tag{B.7}$$

The resulting equation is a sixth order equation that can be numerically solved giving the optimum value of c_1, and therefore, c_2. It can be shown that there is one and only solution to Eq. (B.7) such that c_1 lies between 0 and c and that the point of inflection corresponds to a minimum.

We note that the system shown in Fig. B.1 is the simplest possible construct of a network. A generalized system (or a network) will have the appearance of multiple nodes and multiple instances of connectivity, along with more than two classes of traffic. A closed form solution to a generalized system, while not impossible in principle, is impractical with the current state-of-the art of analytical techniques. In practices, one can, of course, have varying architectures of networks and launch identical traffic on each and simulate the outcome in terms of pre-determined parameters of performance. This can lead to identifying the desired composition of the optimal network for a pre-determined traffic matrix.

In the following analysis, we try to gain an insight into the larger question of the impact of heterogeneity by taking an oblique approach. More specifically, we ask the following question: If the traffic is heterogeneous, are we better off separating it into two transmission system while keeping the overall network resources (bandwidth) constant, or would an integrated network with combined resources result in better utilization of resources? In other words, at which point, specifically, does the advantage accruing from the economy of scale yield its benefits to dividing the traffic in two (or more) classes and diverting each class to a different transmission system? This larger question is far from being an academic one. For example, if a service provider has access to a satellite network, should it off-load its streaming traffic on to the satellite network and utilize its terrestrial transmission network for largely interactive traffic? The analysis presented below offers an insight that can lead to a commercially driven optimal decision.

We now address the case when the entire traffic is handled by a common pooled resource with a transmission capacity of c where $c = c_1 + c_2$. We term the new system the integrated system. The integrated system will have the following parameters:

- The arrival process will be a Poisson distribution with parameter $\lambda = \lambda_1 + \lambda_2$
- The mean service time will be given by $\frac{1}{\mu c} = \frac{\lambda_1}{\lambda} \frac{1}{\mu_1 c} + \frac{\lambda_2}{\lambda} \frac{1}{\mu_2 c}$
- The variance of the service time can be shown to be given as

$$\sigma_{int}^2 = \frac{1}{c^2} \left[\frac{\lambda_1}{\lambda} \frac{2 - \mu_1}{\mu_1^2} + \frac{\lambda_2}{\lambda} \frac{2 - \mu_2}{\mu_2^2} - \left(\frac{\mu_1}{\lambda \mu_1} + \frac{\lambda_2}{\lambda \mu_2} \right)^2 \right] \tag{B.8}$$

The mean delay per message for the integrated system can now be written as [2]

$$T_{int} = \frac{1}{\mu c} + \frac{\left(\frac{\lambda}{\mu c} \right) + \sigma_{int}^2 \lambda \mu c}{2(\mu c - \lambda)} \tag{B.9}$$

The benefits of integration reduce as the disparity in traffic increases. On the one hand, the economy of scale results in improving the benefits of integration. Quite on the contrary, the disadvantages of integration increase dramatically as the traffic disparity increases.

One way to visualize the impact of integration vs. segregation is as follows.

Assume numerical values of an integrated system with defined values of $1/\mu$, λ and c, ensuring that the utilization factor $\rho = \frac{\lambda}{\mu c}$ is less than 1. Then compute the value of the mean message delay from Eq. (B.1). Keep the value of ρ constant in a corresponding segregated system for a compelling insight into integration vs. segregated systems.

In order to segregate the above integrated system into two comparable systems, proceed as follows.

Choose values of $\frac{\mu_1}{\mu_2}$ over a range of, say, 1–100. The ratio $\frac{\mu_1}{\mu_2}$ is an indicator of disparity and will be used as a parameter in a plot that will illustrate the impact of disparity in a comparative evaluation of integrated and segregated systems.

The overall utilization of the segregated system will be given by

$$\rho_{seg} = \frac{1}{c}\left[\frac{\lambda_1}{\lambda}\frac{1}{\mu_1} + \frac{\lambda_2}{\lambda}\frac{1}{\mu_2}\right] \tag{B.10}$$

Since $\rho_{seg} = \rho$, and the total capacity c is fixed, if we keep the total number of messages constant, i.e., $\lambda = \lambda_1 + \lambda_2$, then Eq. (B.10) can be solved to evaluate λ_1 and λ_2.

In order to split the total resources c into c_1 and c_2 optimally, use Eqs. (B.6) and (B.7). The resulting sixth order equation can be numerically solved under the appropriate boundary conditions, i.e.,

- $c = c_1 + c_2$, and,
- c_1 and c_2 are positive and real.

A plot of the ratios of the lengths of the two segregated systems against the mean delays of integrated and segregated systems will visually illustrate the impacts of segregation and integration. For relatively low ratios of $\frac{\mu_1}{\mu_2}$, i.e., low disparity, the integrated system will have a lower mean delay. As the traffic diversity increases, the mean delay of the integrated system will increase indefinitely.

We draw on previously published results to show that the benefits of integration are bounded, whereas the penalties of integration (in the face of disparate traffic served by an equivalent amount of transmission resource) are unbounded. This is described by the following theorem.

Theorem B.1 *If the traffic generated by the N classes of users has the same mean and variance, then, using the same total channel capacity, the average delay per message for the segregated system is N times the average delay for the integrated system [1].*

Proof We define the following parameters for the systems: $\frac{1}{\mu}$ = mean message length, σ_1^2 = variance of message lengths, λ = arrival rate for each of the N classes of users, c = total channel capacity, $\frac{c}{N}$ = channel capacity of each segregated channel. It can be shown that

$$\sigma_{seg}^2 = \frac{N^2\sigma_1^2}{c^2} \tag{B.11}$$

and,

$$\sigma_{int}^2 = \frac{1}{c^2}\left[\frac{1}{N}\left(N\sigma_1^2 + \frac{N}{\mu^2}\right) - \frac{1}{\mu^2}\right] = \frac{\sigma^2}{c^2} \tag{B.12}$$

For the segregated system, the transmission resource is divided by a factor of N, while the incident traffic is lower by a factor of N. Thus the resource utilization factor for both the segregated and integrated systems is identical, i.e.,

$$\rho_{int} = \rho_{seg} \tag{B.13}$$

Using Eqs. (B.3) and (B.9), and substituting the appropriate values for the transmission resources, it can be shown that

$$T_{seg} = \frac{1}{\mu \frac{c}{N}} + \frac{\frac{\rho^2 + \lambda^2 N^2 \sigma_1^2}{c^2}}{2\lambda(1 - \rho)} \tag{B.14}$$

and,

$$T_{int} = \frac{1}{\mu c} + \frac{\frac{\rho^2 + \lambda^2 N^2 \sigma_1^2}{c^2}}{2\lambda(1 - \rho)N} \tag{B.15}$$

Comparing Eqs. (B.14) and (B.15), it is immediately obvious that

$$\frac{T_{seg}}{T_{int}} = N \tag{B.16}$$

We have thus proven that an integrated system with N identical traffic classes has lower average delay by a factor of N compared to a segregated system with the total traffic divided into N systems, with each system allocated a bandwidth resource of $\frac{c}{N}$.

It can be shown that with just two classes of traffic with widely diverse characteristics in terms of mean message lengths, even with the overall utilization factor remaining constant, there is an indefinite increase in average traffic delay as the disparity increases [1]. This leads us to conclude that the benefits of integration are bounded, while the penalties associated with integration in the face of disparate traffic classes are unbounded. As stated, the performance parameter in terms of which the benefit or the penalty is identified is the mean delay associated with traffic.

This observation is also intuitive for transportation networks. In the Federal Republic of Germany on Autobahns, traffic lanes are segregated with minimum speed limits. In department stores in the USA, separate checkout counters are often provided to customers who have no more than a specified number of items in their shopping cart which corresponds to lower service time for such customers.

The reader will recognize the fact that merger of networks is also governed by the strategic intents of the involved entities. It is beyond the scope of this book to elaborate on the business considerations related to merger of networks. These considerations cannot be easily captured in analytical results. Resource considerations are, of course, an important consideration. Merger considerations based on overall resource consideration can be quantified and captured in the approaches discussed above.

Telecommunication networks are dynamic entities. The evolution of traffic and its characteristics, quality of service anticipated by the community of users, evolving price-performance characteristics of the fundamental building blocks of the network, and the evolving laws that regulate the business of telecommunications, among others, profoundly affect considerations related to merger of networks.

References

1. P.K. Verma (ed.), *ISDN Sysrtems—Architecture, Technology, and Applications* (Prentice Hall, Englewood Cliffs, 1990), pp. 61–70
2. D.G. Kendall, Some problems in the theory of queues. J. Roy. Statis. Soc. Series B **13**, 151–185 (1951)

Bibliography

1. S. Blake, D. Black, M. Carhn, E. Daviq, Z. Wang, W. Weiss, *An Architecture for Differentiated Servicss, RFC2475* (1998)
2. F. Zhang, P. Verma, S. Cheng, Pricing, resource allocation and QoS in multi-class networks with competitive market model. IET Commun. **5**(1), 51–60 (2011). https://doi.org/10.1049/iet-com.2009.0694
3. L. Walras, *Elements of Pure Economics; Or the Theory of Social Wealth* (Lausanne, Paris, 1874)
4. K.J. Arrow, G. Debreu, Existence of an equilibrium for a competitive economy. Econo-metrica **22**(3), 265–290 (1954)
5. W.C. Brainard, H. Scarf, How to compute equilibrium prices in 1891, in *Cowles Foundation Discussion Paper 1270* (2000)
6. E. Eisenberg, D. Gale, Consensus of subjective probabilities: the pari-mutuel method. Ann. Math. Stat. **30**, 165–168 (1959)
7. E. Eisenberg, Aggregation of utility functions. Manag. Sci. **7**(4), 337–350 (1961)
8. D. Gale, *The Theory of Linear Economic Models* (McGraw Hill, New York, 1960)
9. M. Ling, J. Tsai, Y. Ye, Budget allocation in a competitve communication spectrum economy. EURASIP J. Adv. Signal Process. **2009**(1), 963717 (2009)
10. W. Rudin, *Principles of Mathematical Analysis*, 3rd edn. (McGraw-Hill, New York, 1976), p 101
11. L. Kleinrock, Communication nets: stochastic message flow and delay (McGraw-Hill, New York, 1994)
12. L. Chen, Y. Ye, J. Zhang, A note on equilibrium pricing as convex optimization, Working Paper (2007)
13. H.R. Sukasdadi, P.K. Verma, A constant revenue model for telecommunication networks, in *International Conference on Systems and International Conference on Mobile Communications and Learning Technologies, ICN/ICONS/MCL* (2006)
14. ITU-T, *One-way Transmission Time, Recommendation G.114* (1996)
15. A. Odlyzko, *Internet Pricing and History of Communications, AT&T Labs- Research* (2001)

© Springer Nature Switzerland AG 2020

P. Verma, F. Zhang, *The Economics of Telecommunication Services*, Textbooks in Telecommunication Engineering, https://doi.org/10.1007/978-3-030-33865-7

Index

© Springer Nature Switzerland AG 2020   191
P. Verma, F. Zhang, *The Economics of Telecommunication Services*, Textbooks in
Telecommunication Engineering, https://doi.org/10.1007/978-3-030-33865-7

Printed in the United States
by Baker & Taylor Publisher Services